TRACTS FOR COMPUTERS
No. XXX

TABLE OF d' AND β

BY

P. R. FREEMAN
University College, London

CAMBRIDGE
AT THE UNIVERSITY PRESS
1973

CAMBRIDGE UNIVERSITY PRESS
Cambridge, New York, Melbourne, Madrid, Cape Town,
Singapore, São Paulo, Delhi, Tokyo, Mexico City

Cambridge University Press
The Edinburgh Building, Cambridge CB2 8RU, UK

Published in the United States of America by Cambridge University Press, New York

www.cambridge.org
Information on this title: www.cambridge.org/9780521294638

First published 1973
First paperback edition 2011

A catalogue record for this publication is available from the British Library

Library of Congress Catalogue Card Number: 72-91360

ISBN 978-0-521-20018-9 Hardback
ISBN 978-0-521-29463-8 Paperback

TABLE OF d' AND β

INTRODUCTION

For a standardised normal distribution (fig. 1) with density function

$$\phi(t) = \frac{1}{\sqrt{(2\pi)}} \exp\left(-\tfrac{1}{2}t^2\right)$$

and distribution function

$$\Phi(t) = \int_{-\infty}^{t} \phi(u)\, du$$

the z-score corresponding to any number p, $0 < p < 1$, is defined by

$$p = \Phi(z) \quad \text{or} \quad z = \Phi^{-1}(p).$$

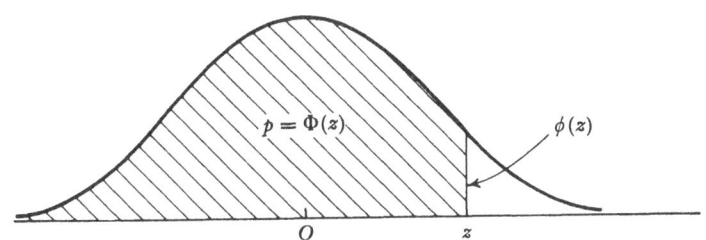

Fig. 1

For a pair of proportions, $p_1 = $ 'false positive rate' and $p_2 = $ 'hit rate', we tabulate

$$d' = z_2 - z_1$$

$$\beta = \frac{\phi(z_2)}{\phi(z_1)} = \exp\left\{\tfrac{1}{2}(z_1^2 - z_2^2)\right\}.$$

In this table, the false positive rate p_1 is shown at the top of each column, and the hit rate p_2 at the left hand side of each row. Each cell of the table contains two entries, the first of which is d' and the second is β. The values of p_1 and p_2 covered are

$p_1 = 0.01\ (0.01)\ 0.30\ (0.005)\ 0.70\ (0.01)\ 0.99\ 0.999$

$p_2 = 0.0001\ (0.0003)\ 0.001\ (0.001)\ 0.01\ (0.002)\ 0.10\ (0.01)\ 0.30\ (0.05)$

$\qquad\qquad 0.40\ (0.10)\ 0.80.$

The table is arranged into fourteen groups, each of 7 consecutive pages, within which the values of p_1 remain fixed while p_2 ranges over its 140 possible values. The single row of values at the bottom of each page labelled **Cutoff** gives $-z_1$, the negative of the z-scores corresponding to the p_1 values at the top of the page. The z_2 corresponding to any hit rate p_2 may be obtained by subtracting cutoff in any column from the appropriate d' value in that column. Values of z corresponding to each value of p, accurate to ten decimal places, were taken from Karl Pearson's *Tables for Statisticians and Biometricians*, Part II, where available, and specially calculated to at least twelve places for $p = 0.0001$ (0.0003) 0.001. These are sufficient to ensure that all tabulated values of d' and β are accurate to the three decimal places given.

Simple linear interpolation should be accurate enough in most parts of the table, but care may be needed where either p_1 or p_2 is close to 0 or 1.

USE OF THE TABLE

The following section will be placed entirely in terms of the theory of signal detection, although the table can, of course, be used in several other contexts.

A fundamental assumption underlying the whole theory is that observers are able to distinguish between various sensory events. Apart from certain distribution-free models, it is also necessary to further assume that these sensory events can be mapped into a single dimension, indexed by a variable x, say.

It is possible to distinguish four experimental situations.

(a) Yes–no experiments

Here an observer is presented with a sequence of trials, at each of which he is required to say whether or not a signal was present. The data can be summarised as

		Signal present	Signal absent
Reported $\begin{cases} \\ \\ \end{cases}$	Signal present	a	b
	Signal absent	c	d

from which

$$p_1 = \text{false positive rate} = \frac{b}{b+d},$$

$$p_2 = \text{hit rate} = \frac{a}{a+c}.$$

The basic model (see fig. 2) assumes that when a signal is present, a value of the variable x is generated according to a normal distribution with mean μ_s and variance σ^2, while when a signal is absent, the normal distribution has lower mean μ_n, but the same variance σ^2. The observer is assumed to determine a cutoff value c such that if he perceives a value of x greater than c he reports a signal, while if x is less than c he reports no signal.

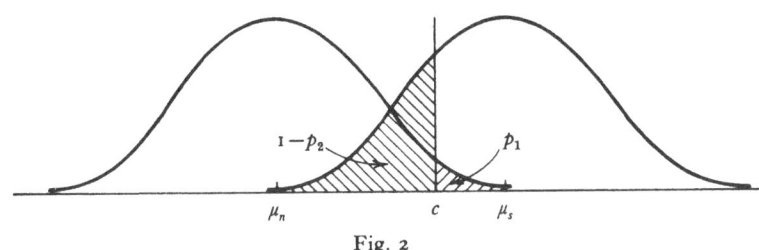

Fig. 2

We then have $d' = (\mu_s - \mu_n)/\sigma$, the standardised difference between the means and

$$\beta = \text{likelihood ratio at the cutoff value } c.$$

The table can be entered with the observed values of p_1 and p_2 to yield estimates of d', β and at the bottom of each page, $(c - \mu_n)/\sigma$ is given by cutoff.

If the experimental conditions are manipulated so as to obtain several pairs of values of p_1, p_2 from the same observer, a plot of p_2 against p_1 yields the so-called ROC curve, as in fig. 3. Equivalently, a plot of the z-scores, z_2 against z_1 should, if all the assumptions hold, yield a straight line of slope 1 and intercept d'. If the line is not straight, one can only conclude that the model does not fit the data. If, however, it is straight but has slope other than 1, this indicates that the model holds except for the assumption of equal variances. Denoting the variances of the signal and noise distributions by σ_s^2 and σ_n^2, respectively, the slope of the line is σ_n/σ_s.

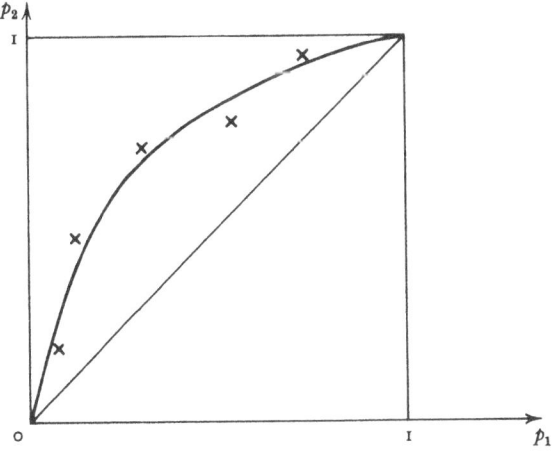

Fig. 3

In order to proceed, it is convenient to assume that the difference in the means of the two distributions is proportional to the change either in standard deviation or in variance. In the first case, let $\Delta m/\Delta \sigma = \nu$, then

$$z_2 = \left(1+\frac{d'}{\nu}\right)^{-1}(z_1+d')$$

where now $d' = (\mu_s - \mu_n)/\sigma_n$. In the second case, let $\Delta m/\Delta \sigma^2 = \nu^1$, then

$$z_2 = \left(1+\frac{d'}{\nu^1 \sigma_n}\right)^{-\frac{1}{2}}(z_1+d').$$

The two cases are therefore logically indistinguishable. The value of d' may be determined as the negative intercept on the z_1 axis, and the slope of the line used to estimate either ν or $\nu^1 \sigma_n$, whichever case is preferred. See fig. 4.

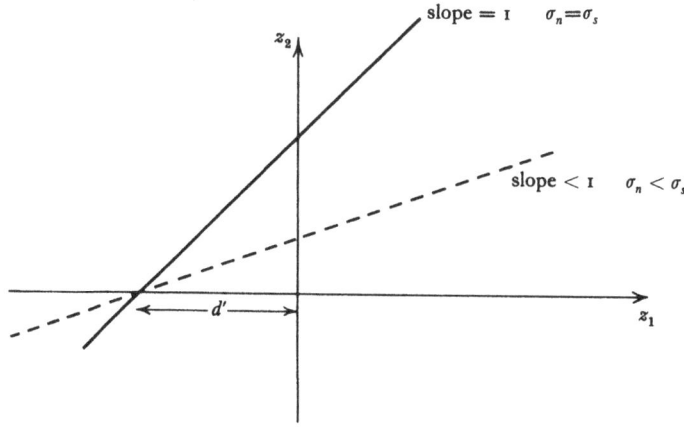

Fig. 4

(b) Rating scale experiments

Here the observer is presented with a sequence of trials and is required to say for each trial whether or not a signal was presented, and if so an indication is required of how confident the observer is of his response. This may be achieved by providing k discrete categories of increasing confidence or a continuous sliding scale. An example of such data might be

		Signal present	Signal absent
Reported signal	Most confident	a_1	b_1
	Very confident	a_2	b_2
	...	...	...
	Rather doubtful	a_k	b_k
	Signal absent	c	d

Separate values of d' and β may be obtained by entering the table with

$$p_1 = \frac{b_1}{B+d}, \quad p_2 = \frac{a_1}{A+c},$$

$$p_1 = \frac{b_1+b_2}{B+d}, \quad p_2 = \frac{a_1+a_2}{A+c}, \quad \cdots$$

$$p_1 = \frac{B}{B+d}, \quad p_2 = \frac{A}{A+c},$$

where $A = \sum_1^k a_i$, $B = \sum_1^k b_i$.

The model is the same as that for the yes–no experiment, it is assumed that the observer generates the various confidence ratings by using a succession of cutoffs $c_1, \ldots, c_k$. The same considerations apply about estimating d' and checking the assumption of equal variances.

(c) Two-alternative forced choice experiments

Here, at each trial, the observer is given two stimulus presentations and asked to report which one he thinks contained the signal. The data again takes the form

		$S1$: Signal in first stimulus	$S2$: Signal in second stimulus
Reported	$R1$: signal in first stimulus	a	b
	$R2$: signal in second stimulus	c	d

and
$$p_1 = p(R1|S2) = \frac{b}{b+d}, \quad p_2 = p(R1|S1) = \frac{a}{a+c}.$$

The model assumes that the observer receives values x_1 from the first stimulus and x_2 from the second, where given $S1$, x_1 is normal mean μ_s, variance σ_s^2 and x_2 is normal mean μ_n, variance σ_n^2 while, given $S2$ the means and variances are interchanged. In either case there is no need to assume independence of x_1 and x_2. Let r denote the correlation between them. It is assumed that the observer responds $R1$ if $x_1 - x_2 > c$ and $R2$ if $x_1 - x_2 < c$, so that

$$p(R1|S1) = 1 - \Phi\left(\frac{c - \mu_s + \mu_n}{\sigma}\right),$$

$$p(R1|S2) = 1 - \Phi\left(\frac{c - \mu_n + \mu_s}{\sigma}\right),$$

where $\sigma^2 = \sigma_s^2 + \sigma_n^2 - 2r\sigma_s\sigma_n$.

It is thus not necessary to assume $\sigma_s = \sigma_n$ here. The table can be entered with the observed values of p_1 and p_2 to determine $d' = 2(m_s - m_n)/\sigma$, β and c. Note that *if* $\sigma_s = \sigma_n$ *and* $r = 0$ then $\sigma = \sigma_n \sqrt{2}$, so this observed value of d' should be divided by $\sqrt{2}$ to make it comparable with a value obtained from a yes–no experiment.

(d) m-alternative forced choice experiments

The above model for two-alternative experiments clearly will not generalise easily. It is commonly replaced by the following, of which the above model with $c = 0$ is a special case. The observer is assumed to receive independent values $x_1, \dots, x_m$ from the m stimuli, one of which is from a normal distribution with mean μ_s, variance σ^2 and all the others are from a normal distribution with mean μ_n and the same variance σ^2. The observer then responds that the stimulus which yielded the largest value of x is the one that contained the signal.

It follows that the proportion of correct responses is

$$p(C) = \int_{-\infty}^{\infty} \phi(x - d') \, [\Phi(x)]^{m-1} \, \mathrm{d}x$$

where $d' = (\mu_s - \mu_n)/\sigma$.

Tables in *Signal detection and recognition by human observers* edited by J. A. Swets (1964) permit estimation of d' from observed values of $p(C)$.

Hit Rate	0·0001	0·0004	0·0007	0·0010	0·0020	0·0030
0·010	1·393	1·026	0·868	0·764	0·552	0·421
	67·329	18·443	10·990	7·915	4·204	2·913
0·020	1·665	1·299	1·141	1·036	0·824	0·694
	122·315	33·504	19·964	14·380	7·637	5·292
0·030	1·838	1·472	1·314	1·209	0·997	0·867
	171·889	47·084	28·056	20·208	10·732	7·437
0·040	1·968	1·602	1·444	1·340	1·127	0·997
	217·694	59·630	35·532	25·593	13·592	9·419
0·050	2·074	1·708	1·550	1·445	1·233	1·103
	260·544	71·368	42·526	30·630	16·267	11·273
0·060	2·164	1·798	1·640	1·535	1·323	1·193
	300·931	82·431	49·119	35·379	18·789	13·020
0·070	2·243	1·877	1·719	1·614	1·402	1·272
	339·191	92·911	55·363	39·877	21·177	14·675
0·080	2·314	1·948	1·790	1·685	1·473	1·343
	375·564	102·874	61·300	44·153	23·448	16·249
0·090	2·378	2·012	1·854	1·749	1·537	1·407
	410·235	112·371	66·959	48·229	25·613	17·749
0·100	2·437	2·071	1·913	1·809	1·597	1·466
	443·348	121·441	72·364	52·122	27·680	19·182
0·110	2·492	2·126	1·968	1·864	1·652	1·521
	475·020	130·117	77·534	55·845	29·658	20·552
0·120	2·544	2·178	2·020	1·915	1·703	1·573
	505·347	138·424	82·484	59·410	31·551	21·864
0·130	2·593	2·226	2·068	1·964	1·752	1·621
	534·410	146·385	87·227	62·827	33·366	23·122
0·140	2·639	2·272	2·114	2·010	1·798	1·667
	562·278	154·018	91·776	66·104	35·106	24·327
0·150	2·683	2·316	2·158	2·054	1·842	1·711
	589·011	161·341	96·140	69·246	36·775	25·484
0·160	2·725	2·358	2·200	2·096	1·884	1·753
	614·660	168·367	100·326	72·262	38·376	26·594
0·170	2·765	2·399	2·240	2·136	1·924	1·794
	639·270	175·108	104·343	75·155	39·913	27·659
0·180	2·804	2·437	2·279	2·175	1·963	1·832
	662·881	181·575	108·197	77·931	41·387	28·680
0·190	2·841	2·475	2·317	2·212	2·000	1·870
	685·529	187·779	111·893	80·593	42·801	29·660
0·200	2·877	2·511	2·353	2·249	2·037	1·906
	707·246	193·728	115·438	83·147	44·157	30·600
Cutoff	3·719	3·353	3·195	3·090	2·878	2·748

Hit Rate	0·0001	0·0004	0·0007	0·0010	0·0020	0·0030
0·210	2·913 728·061	2·546 199·429	2·388 118·835	2·284 85·594	2·072 45·456	1·941 31·500
0·220	2·947 747·998	2·581 204·890	2·422 122·090	2·318 87·938	2·106 46·701	1·976 32·363
0·230	2·980 767·083	2·614 210·118	2·456 125·205	2·351 90·181	2·139 47·893	2·009 33·189
0·240	3·013 785·335	2·646 215·118	2·488 128·184	2·384 92·327	2·172 49·032	2·041 33·978
0·250	3·045 802·774	2·678 219·895	2·520 131·030	2·416 94·377	2·204 50·121	2·073 34·733
0·260	3·076 819·419	2·709 224·454	2·551 133·747	2·447 96·334	2·235 51·160	2·104 35·453
0·270	3·106 835·284	2·740 228·800	2·582 136·337	2·477 98·199	2·265 52·151	2·135 36·139
0·280	3·136 850·385	2·770 232·936	2·612 138·802	2·507 99·975	2·295 53·094	2·165 36·793
0·290	3·166 864·736	2·799 236·867	2·641 141·144	2·537 101·662	2·325 53·990	2·194 37·414
0·300	3·195 878·349	2·828 240·596	2·670 143·366	2·566 103·262	2·354 54·839	2·223 38·003
0·305	3·209 884·882	2·843 242·385	2·685 144·432	2·580 104·030	2·368 55·247	2·238 38·285
0·310	3·223 891·235	2·857 244·126	2·699 145·469	2·594 104·777	2·382 55·644	2·252 38·560
0·315	3·237 897·409	2·871 245·817	2·713 146·477	2·609 105·503	2·396 56·029	2·266 38·827
0·320	3·251 903·405	2·885 247·459	2·727 147·456	2·623 106·208	2·410 56·404	2·280 39·087
0·325	3·265 909·224	2·899 249·053	2·741 148·405	2·636 106·892	2·424 56·767	2·294 39·338
0·330	3·279 914·868	2·913 250·599	2·755 149·327	2·650 107·555	2·438 57·120	2·308 39·583
0·335	3·293 920·338	2·927 252·097	2·769 150·219	2·664 108·198	2·452 57·461	2·322 39·819
0·340	3·307 925·634	2·940 253·548	2·782 151·084	2·678 108·821	2·466 57·792	2·335 40·048
0·345	3·320 930·758	2·954 254·952	2·796 151·920	2·691 109·423	2·479 58·112	2·349 40·270
0·350	3·334 935·710	2·967 256·308	2·809 152·728	2·705 110·006	2·493 58·421	2·362 40·484
Cutoff	3·719	3·353	3·195	3·090	2·878	2·748

Hit Rate	0·0001	0·0004	0·0007	0·0010	0·0020	0·0030
0·355	3·347 940·492	2·981 257·618	2·823 153·509	2·718 110·568	2·506 58·719	2·376 40·691
0·360	3·361 945·104	2·994 258·881	2·836 154·262	2·732 111·110	2·520 59·007	2·389 40·891
0·365	3·374 949·548	3·008 260·099	2·850 154·987	2·745 111·632	2·533 59·285	2·403 41·083
0·370	3·387 953·823	3·021 261·270	2·863 155·685	2·758 112·135	2·546 59·552	2·416 41·268
0·375	3·400 957·932	3·034 262·395	2·876 156·355	2·772 112·618	2·560 59·808	2·429 41·446
0·380	3·414 961·873	3·047 263·475	2·889 156·999	2·785 113·081	2·573 60·054	2·442 41·616
0·385	3·427 965·649	3·060 264·509	2·902 157·615	2·798 113·525	2·586 60·290	2·455 41·780
0·390	3·440 969·259	3·073 265·498	2·915 158·204	2·811 113·950	2·599 60·515	2·468 41·936
0·395	3·453 972·705	3·086 266·442	2·928 158·767	2·824 114·355	2·612 60·731	2·481 42·085
0·400	3·466 975·987	3·099 267·341	2·941 159·303	2·837 114·741	2·625 60·935	2·494 42·227
0·405	3·479 979·106	3·112 268·195	2·954 159·812	2·850 115·107	2·638 61·130	2·507 42·362
0·410	3·491 982·061	3·125 269·004	2·967 160·294	2·863 115·455	2·651 61·315	2·520 42·490
0·415	3·504 984·854	3·138 269·770	2·980 160·750	2·876 115·783	2·663 61·489	2·533 42·611
0·420	3·517 987·485	3·151 270·490	2·993 161·179	2·888 116·093	2·676 61·653	2·546 42·724
0·425	3·530 989·954	3·164 271·167	3·006 161·582	2·901 116·383	2·689 61·807	2·559 42·831
0·430	3·543 992·263	3·176 271·799	3·018 161·959	2·914 116·654	2·702 61·952	2·571 42·931
0·435	3·555 994·410	3·189 272·387	3·031 162·310	2·927 116·907	2·715 62·086	2·584 43·024
0·440	3·568 996·397	3·202 272·931	3·044 162·634	2·939 117·140	2·727 62·210	2·597 43·110
0·445	3·581 998·224	3·214 273·432	3·056 162·932	2·952 117·355	2·740 62·324	2·609 43·189
0·450	3·593 999·891	3·227 273·888	3·069 163·204	2·965 117·551	2·753 62·428	2·622 43·261
Cutoff	3·719	3·353	3·195	3·090	2·878	2·748

Hit Rate	0·0001	0·0004	0·0007	0·0010	0·0020	0·0030
0·455	3·606 1001·399	3·240 274·301	3·082 163·450	2·977 117·728	2·765 62·522	2·635 43·326
0·460	3·619 1002·747	3·252 274·671	3·094 163·670	2·990 117·887	2·778 62·606	2·647 43·385
0·465	3·631 1003·936	3·265 274·996	3·107 163·864	3·002 118·027	2·790 62·680	2·660 43·436
0·470	3·644 1004·966	3·278 275·279	3·119 164·033	3·015 118·148	2·803 62·745	2·673 43·481
0·475	3·656 1005·837	3·290 275·517	3·132 164·175	3·028 118·250	2·815 62·799	2·685 43·518
0·480	3·669 1006·550	3·303 275·713	3·144 164·291	3·040 118·334	2·828 62·844	2·698 43·549
0·485	3·681 1007·104	3·315 275·864	3·157 164·382	3·053 118·399	2·841 62·878	2·710 43·573
0·490	3·694 1007·500	3·328 275·973	3·170 164·446	3·065 118·446	2·853 62·903	2·723 43·590
0·495	3·706 1007·738	3·340 276·038	3·182 164·485	3·078 118·474	2·866 62·918	2·735 43·601
0·500	3·719 1007·817	3·353 276·059	3·195 164·498	3·090 118·483	2·878 62·923	2·748 43·604
0·505	3·732 1007·738	3·365 276·038	3·207 164·485	3·103 118·474	2·891 62·918	2·760 43·601
0·510	3·744 1007·500	3·378 275·973	3·220 164·446	3·115 118·446	2·903 62·903	2·773 43·590
0·515	3·757 1007·104	3·390 275·864	3·232 164·382	3·128 118·399	2·916 62·878	2·785 43·573
0·520	3·769 1006·550	3·403 275·713	3·245 164·291	3·140 118·334	2·928 62·844	2·798 43·549
0·525	3·782 1005·837	3·416 275·517	3·257 164·175	3·153 118·250	2·941 62·799	2·810 43·518
0·530	3·794 1004·966	3·428 275·279	3·270 164·033	3·166 118·148	2·953 62·745	2·823 43·481
0·535	3·807 1003·936	3·441 274·996	3·282 163·864	3·178 118·027	2·966 62·680	2·836 43·436
0·540	3·819 1002·747	3·453 274·671	3·295 163·670	3·191 117·887	2·979 62·606	2·848 43·385
0·545	3·832 1001·399	3·466 274·301	3·308 163·450	3·203 117·728	2·991 62·522	2·861 43·326
0·550	3·845 999·891	3·478 273·888	3·320 163·204	3·216 117·551	3·004 62·428	2·873 43·261
Cutoff	**3·719**	**3·353**	**3·195**	**3·090**	**2·878**	**2·748**

Hit Rate	0·0001	0·0004	0·0007	0·0010	0·0020	0·0030
0·555	3·857 998·224	3·491 273·432	3·333 162·932	3·229 117·355	3·016 62·324	2·886 43·189
0·560	3·870 996·397	3·504 272·931	3·346 162·634	3·241 117·140	3·029 62·210	2·899 43·110
0·565	3·883 994·410	3·516 272·387	3·358 162·310	3·254 116·907	3·042 62·086	2·911 43·024
0·570	3·895 992·263	3·529 271·799	3·371 161·959	3·267 116·654	3·055 61·952	2·924 42·931
0·575	3·908 989·954	3·542 271·167	3·384 161·582	3·279 116·383	3·067 61·807	2·937 42·831
0·580	3·921 987·485	3·555 270·490	3·397 161·179	3·292 116·093	3·080 61·653	2·950 42·724
0·585	3·934 984·854	3·567 269·770	3·409 160·750	3·305 115·783	3·093 61·489	2·962 42·611
0·590	3·947 982·061	3·580 269·004	3·422 160·294	3·318 115·455	3·106 61·315	2·975 42·490
0·595	3·959 979·106	3·593 268·195	3·435 159·812	3·331 115·107	3·119 61·130	2·988 42·362
0·600	3·972 975·987	3·606 267·341	3·448 159·303	3·344 114·741	3·132 60·935	3·001 42·227
0·605	3·985 972·705	3·619 266·442	3·461 158·767	3·357 114·355	3·144 60·731	3·014 42·085
0·610	3·998 969·259	3·632 265·498	3·474 158·204	3·370 113·950	3·157 60·515	3·027 41·936
0·615	4·011 965·649	3·645 264·509	3·487 157·615	3·383 113·525	3·171 60·290	3·040 41·780
0·620	4·024 961·873	3·658 263·475	3·500 156·999	3·396 113·081	3·184 60·054	3·053 41·616
0·625	4·038 957·932	3·671 262·395	3·513 156·355	3·409 112·618	3·197 59·808	3·066 41·446
0·630	4·051 953·823	3·685 261·270	3·527 155·685	3·422 112·135	3·210 59·552	3·080 41·268
0·635	4·064 949·548	3·698 260·099	3·540 154·987	3·435 111·632	3·223 59·285	3·093 41·083
0·640	4·077 945·104	3·711 258·881	3·553 154·262	3·449 111·110	3·237 59·007	3·106 40·891
0·645	4·091 940·492	3·725 257·618	3·567 153·509	3·462 110·568	3·250 58·719	3·120 40·691
0·650	4·104 935·710	3·738 256·308	3·580 152·728	3·476 110·006	3·263 58·421	3·133 40·484
Cutoff	3·719	3·353	3·195	3·090	2·878	2·748

Hit Rate	0·0001	0·0004	0·0007	0·0010	0·0020	0·0030
0·655	4·118 930·758	3·752 254·952	3·594 151·920	3·489 109·423	3·277 58·112	3·147 40·270
0·660	4·131 925·634	3·765 253·548	3·607 151·084	3·503 108·821	3·291 57·792	3·160 40·048
0·665	4·145 920·338	3·779 252·097	3·621 150·219	3·516 108·198	3·304 57·461	3·174 39·819
0·670	4·159 914·868	3·793 250·599	3·635 149·327	3·530 107·555	3·318 57·120	3·188 39·583
0·675	4·173 909·224	3·807 249·053	3·648 148·405	3·544 106·892	3·332 56·767	3·202 39·338
0·680	4·187 903·405	3·820 247·459	3·662 147·456	3·558 106·208	3·346 56·404	3·215 39·087
0·685	4·201 897·409	3·835 245·817	3·676 146·477	3·572 105·503	3·360 56·029	3·230 38·827
0·690	4·215 891·235	3·849 244·126	3·691 145·469	3·586 104·777	3·374 55·644	3·244 38·560
0·695	4·229 884·882	3·863 242·385	3·705 144·432	3·600 104·030	3·388 55·247	3·258 38·285
0·700	4·243 878·349	3·877 240·596	3·719 143·366	3·615 103·262	3·403 54·839	3·272 38·003
0·710	4·272 864·736	3·906 236·867	3·748 141·144	3·644 101·662	3·432 53·990	3·301 37·414
0·720	4·302 850·385	3·936 232·936	3·777 138·802	3·673 99·975	3·461 53·094	3·331 36·793
0·730	4·332 835·284	3·966 228·800	3·807 136·337	3·703 98·199	3·491 52·151	3·361 36·139
0·740	4·362 819·419	3·996 224·454	3·838 133·747	3·734 96·334	3·522 51·160	3·391 35·453
0·750	4·394 802·774	4·027 219·895	3·869 131·030	3·765 94·377	3·553 50·121	3·422 34·733
0·760	4·425 785·335	4·059 215·118	3·901 128·184	3·797 92·327	3·584 49·032	3·454 33·978
0·770	4·458 767·083	4·092 210·118	3·933 125·205	3·829 90·181	3·617 47·893	3·487 33·189
0·780	4·491 747·998	4·125 204·890	3·967 122·090	3·862 87·938	3·650 46·701	3·520 32·363
0·790	4·525 728·061	4·159 199·429	4·001 118·835	3·897 85·594	3·685 45·456	3·554 31·500
0·800	4·561 707·246	4·194 193·728	4·036 115·438	3·932 83·147	3·720 44·157	3·589 30·600
Cutoff	3·719	3·353	3·195	3·090	2·878	2·748

Hit Rate	0·0001	0·0004	0·0007	0·0010	0·0020	0·0030
0·810	4·597 685·529	4·231 187·779	4·073 111·893	3·968 80·593	3·756 42·801	3·626 29·660
0·820	4·634 662·881	4·268 181·575	4·110 108·197	4·006 77·931	3·794 41·387	3·663 28·680
0·830	4·673 639·270	4·307 175·108	4·149 104·343	4·044 75·155	3·832 39·913	3·702 27·659
0·840	4·713 614·660	4·347 168·367	4·189 100·326	4·085 72·262	3·873 38·376	3·742 26·594
0·850	4·755 589·011	4·389 161·341	4·231 96·140	4·127 69·246	3·915 36·775	3·784 25·484
0·860	4·799 562·278	4·433 154·018	4·275 91·776	4·171 66·104	3·958 35·106	3·828 24·327
0·870	4·845 534·410	4·479 146·385	4·321 87·227	4·217 62·827	4·005 33·366	3·874 23·122
0·880	4·894 505·347	4·528 138·424	4·370 82·484	4·265 59·410	4·053 31·551	3·923 21·864
0·890	4·946 475·020	4·579 130·117	4·421 77·534	4·317 55·845	4·105 29·658	3·974 20·552
0·900	5·001 443·348	4·634 121·441	4·476 72·364	4·372 52·122	4·160 27·680	4·029 19·182
0·910	5·060 410·235	4·694 112·371	4·535 66·959	4·431 48·229	4·219 25·613	4·089 17·749
0·920	5·124 375·564	4·758 102·874	4·600 61·300	4·495 44·153	4·283 23·448	4·153 16·249
0·930	5·195 339·191	4·829 92·911	4·670 55·363	4·566 39·877	4·354 21·177	4·224 14·675
0·940	5·274 300·931	4·908 82·431	4·749 49·119	4·645 35·379	4·433 18·789	4·303 13·020
0·950	5·364 260·544	4·998 71·368	4·840 42·526	4·735 30·630	4·523 16·267	4·393 11·273
0·960	5·470 217·694	5·103 59·630	4·945 35·532	4·841 25·593	4·629 13·592	4·498 9·419
0·970	5·600 171·889	5·234 47·084	5·075 28·056	4·971 20·208	4·759 10·732	4·629 7·437
0·980	5·773 122·315	5·407 33·504	5·248 19·964	5·144 14·380	4·932 7·637	4·802 5·292
0·990	6·045 67·329	5·679 18·443	5·521 10·990	5·417 7·915	5·205 4·204	5·074 2·913
0·999	6·809 8·506	6·443 2·330	6·285 1·388	6·180 1·000	5·968 0·531	5·838 0·368
Cutoff	3·719	3·353	3·195	3·090	2·878	2·748

Hit Rate	0·0040	0·0050	0·0060	0·0070	0·0080	0·0090
0·010	0·326 2·250	0·249 1·843	0·186 1·568	0·131 1·368	0·083 1·216	0·039 1·097
0·020	0·598 4·087	0·522 3·348	0·458 2·848	0·404 2·485	0·355 2·209	0·312 1·992
0·030	0·771 5·743	0·695 4·706	0·631 4·002	0·576 3·492	0·528 3·104	0·485 2·799
0·040	0·901 7·274	0·825 5·960	0·761 5·068	0·707 4·422	0·658 3·931	0·615 3·545
0·050	1·007 8·706	0·931 7·133	0·867 6·066	0·812 5·293	0·764 4·705	0·721 4·243
0·060	1·097 10·055	1·021 8·238	0·957 7·006	0·902 6·113	0·854 5·435	0·811 4·901
0·070	1·176 11·333	1·100 9·286	1·036 7·897	0·981 6·890	0·933 6·125	0·890 5·524
0·080	1·247 12·549	1·171 10·281	1·107 8·744	1·052 7·629	1·004 6·782	0·961 6·116
0·090	1·311 13·707	1·235 11·231	1·171 9·551	1·117 8·333	1·068 7·408	1·025 6·681
0·100	1·371 14·814	1·294 12·137	1·231 10·322	1·176 9·006	1·127 8·006	1·084 7·220
0·110	1·426 15·872	1·349 13·004	1·286 11·059	1·231 9·649	1·182 8·578	1·139 7·736
0·120	1·477 16·885	1·401 13·834	1·337 11·765	1·282 10·265	1·234 9·126	1·191 8·230
0·130	1·526 17·856	1·449 14·630	1·386 12·442	1·331 10·856	1·283 9·651	1·239 8·703
0·140	1·572 18·788	1·496 15·393	1·432 13·091	1·377 11·422	1·329 10·154	1·285 9·157
0·150	1·616 19·681	1·539 16·125	1·476 13·713	1·421 11·965	1·372 10·637	1·329 9·592
0·160	1·658 20·538	1·581 16·827	1·518 14·310	1·463 12·486	1·414 11·100	1·371 10·010
0·170	1·698 21·360	1·622 17·501	1·558 14·883	1·503 12·986	1·455 11·545	1·411 10·411
0·180	1·737 22·149	1·660 18·147	1·597 15·433	1·542 13·465	1·494 11·971	1·450 10·795
0·190	1·774 22·906	1·698 18·767	1·634 15·960	1·579 13·925	1·531 12·380	1·488 11·164
0·200	1·810 23·631	1·734 19·361	1·671 16·466	1·616 14·367	1·567 12·772	1·524 11·518
Cutoff	2·652	2·576	2·512	2·457	2·409	2·366

Hit Rate	0·0040	0·0050	0·0060	0·0070	0·0080	0·0090
0·210	1·846 24·327	1·769 19·931	1·706 16·950	1·651 14·789	1·602 13·148	1·559 11·857
0·220	1·880 24·993	1·804 20·477	1·740 17·414	1·685 15·194	1·637 13·508	1·593 12·182
0·230	1·913 25·631	1·837 21·000	1·773 17·859	1·718 15·582	1·670 13·853	1·627 12·492
0·240	1·946 26·241	1·870 21·499	1·806 18·284	1·751 15·953	1·703 14·182	1·659 12·790
0·250	1·978 26·823	1·901 21·977	1·838 18·690	1·783 16·307	1·734 14·497	1·691 13·074
0·260	2·009 27·379	1·932 22·432	1·869 19·077	1·814 16·645	1·766 14·798	1·722 13·345
0·270	2·039 27·910	1·963 22·867	1·899 19·446	1·844 16·968	1·796 15·084	1·753 13·603
0·280	2·069 28·414	1·993 23·280	1·929 19·798	1·874 17·274	1·826 15·357	1·783 13·849
0·290	2·099 28·894	2·022 23·673	1·959 20·132	1·904 17·566	1·856 15·616	1·812 14·083
0·300	2·128 29·348	2·051 24·046	1·988 20·449	1·933 17·842	1·885 15·862	1·841 14·305
0·305	2·142 29·567	2·066 24·224	2·002 20·601	1·947 17·975	1·899 15·980	1·856 14·411
0·310	2·156 29·779	2·080 24·398	2·016 20·749	1·961 18·104	1·913 16·095	1·870 14·514
0·315	2·170 29·985	2·094 24·567	2·030 20·893	1·976 18·230	1·927 16·206	1·884 14·615
0·320	2·184 30·186	2·108 24·731	2·044 21·032	1·990 18·351	1·941 16·315	1·898 14·713
0·325	2·198 30·380	2·122 24·891	2·058 21·168	2·004 18·470	1·955 16·420	1·912 14·807
0·330	2·212 30·569	2·136 25·045	2·072 21·299	2·017 18·584	1·969 16·522	1·926 14·899
0·335	2·226 30·751	2·150 25·195	2·086 21·427	2·031 18·695	1·983 16·620	1·939 14·988
0·340	2·240 30·928	2·163 25·340	2·100 21·550	2·045 18·803	1·996 16·716	1·953 15·075
0·345	2·253 31·100	2·177 25·480	2·113 21·669	2·058 18·907	2·010 16·809	1·967 15·158
0·350	2·267 31·265	2·191 25·616	2·127 21·784	2·072 19·008	2·024 16·898	1·980 15·239
Cutoff	2·652	2·576	2·512	2·457	2·409	2·366

Hit Rate	0·0040	0·0050	0·0060	0·0070	0·0080	0·0090
0·355	2·280 31·425	2·204 25·747	2·140 21·896	2·085 19·105	2·037 16·984	1·994 15·317
0·360	2·294 31·579	2·217 25·873	2·154 22·003	2·099 19·198	2·050 17·068	2·007 15·392
0·365	2·307 31·727	2·231 25·995	2·167 22·107	2·112 19·289	2·064 17·148	2·020 15·464
0·370	2·320 31·870	2·244 26·112	2·180 22·206	2·125 19·375	2·077 17·225	2·034 15·534
0·375	2·333 32·008	2·257 26·224	2·194 22·302	2·139 19·459	2·090 17·299	2·047 15·601
0·380	2·347 32·139	2·270 26·332	2·207 22·394	2·152 19·539	2·103 17·371	2·060 15·665
0·385	2·360 32·265	2·283 26·435	2·220 22·482	2·165 19·616	2·117 17·439	2·073 15·726
0·390	2·373 32·386	2·297 26·534	2·233 22·566	2·178 19·689	2·130 17·504	2·086 15·785
0·395	2·386 32·501	2·310 26·629	2·246 22·646	2·191 19·759	2·143 17·566	2·099 15·841
0·400	2·399 32·611	2·322 26·718	2·259 22·722	2·204 19·826	2·156 17·625	2·112 15·895
0·405	2·412 32·715	2·335 26·804	2·272 22·795	2·217 19·889	2·168 17·682	2·125 15·945
0·410	2·425 32·814	2·348 26·885	2·285 22·864	2·230 19·949	2·181 17·735	2·138 15·994
0·415	2·437 32·907	2·361 26·961	2·297 22·929	2·243 20·006	2·194 17·786	2·151 16·039
0·420	2·450 32·995	2·374 27·033	2·310 22·990	2·255 20·059	2·207 17·833	2·164 16·082
0·425	2·463 33·078	2·387 27·101	2·323 23·047	2·268 20·109	2·220 17·878	2·176 16·122
0·430	2·476 33·155	2·399 27·164	2·336 23·101	2·281 20·156	2·233 17·919	2·189 16·160
0·435	2·488 33·226	2·412 27·223	2·348 23·151	2·294 20·200	2·245 17·958	2·202 16·195
0·440	2·501 33·293	2·425 27·277	2·361 23·197	2·306 20·240	2·258 17·994	2·215 16·227
0·445	2·514 33·354	2·438 27·327	2·374 23·240	2·319 20·277	2·271 18·027	2·227 16·257
0·450	2·526 33·410	2·450 27·373	2·386 23·279	2·332 20·311	2·283 18·057	2·240 16·284
Cutoff	2·652	2·576	2·512	2·457	2·409	2·366

Hit Rate	0·0040	0·0050	0·0060	0·0070	0·0080	0·0090
0·455	2·539 33·460	2·463 27·414	2·399 23·314	2·344 20·342	2·296 18·084	2·253 16·308
0·460	2·552 33·505	2·475 27·451	2·412 23·345	2·357 20·369	2·308 18·109	2·265 16·330
0·465	2·564 33·545	2·488 27·484	2·424 23·373	2·369 20·393	2·321 18·130	2·278 16·350
0·470	2·577 33·579	2·501 27·512	2·437 23·397	2·382 20·414	2·334 18·149	2·290 16·367
0·475	2·589 33·608	2·513 27·536	2·449 23·417	2·395 20·432	2·346 18·165	2·303 16·381
0·480	2·602 33·632	2·526 27·555	2·462 23·434	2·407 20·447	2·359 18·177	2·315 16·392
0·485	2·614 33·651	2·538 27·570	2·475 23·447	2·420 20·458	2·371 18·187	2·328 16·401
0·490	2·627 33·664	2·551 27·581	2·487 23·456	2·432 20·466	2·384 18·195	2·341 16·408
0·495	2·640 33·672	2·563 27·588	2·500 23·461	2·445 20·471	2·396 18·199	2·353 16·412
0·500	2·652 33·674	2·576 27·590	2·512 23·463	2·457 20·472	2·409 18·200	2·366 16·413
0·505	2·665 33·672	2·588 27·588	2·525 23·461	2·470 20·471	2·421 18·199	2·378 16·412
0·510	2·677 33·664	2·601 27·581	2·537 23·456	2·482 20·466	2·434 18·195	2·391 16·408
0·515	2·690 33·651	2·613 27·570	2·550 23·447	2·495 20·458	2·447 18·187	2·403 16·401
0·520	2·702 33·632	2·626 27·555	2·562 23·434	2·507 20·447	2·459 18·177	2·416 16·392
0·525	2·715 33·608	2·639 27·536	2·575 23·417	2·520 20·432	2·472 18·165	2·428 16·381
0·530	2·727 33·579	2·651 27·512	2·587 23·397	2·533 20·414	2·484 18·149	2·441 16·367
0·535	2·740 33·545	2·664 27·484	2·600 23·373	2·545 20·393	2·497 18·130	2·453 16·350
0·540	2·753 33·505	2·676 27·451	2·613 23·345	2·558 20·369	2·509 18·109	2·466 16·330
0·545	2·765 33·460	2·689 27·414	2·625 23·314	2·570 20·342	2·522 18·084	2·479 16·308
0·550	2·778 33·410	2·701 27·373	2·638 23·279	2·583 20·311	2·535 18·057	2·491 16·284
Cutoff	2·652	2·576	2·512	2·457	2·409	2·366

Hit Rate	0·0040	0·0050	0·0060	0·0070	0·0080	0·0090
0·555	2·790 33·354	2·714 27·327	2·650 23·240	2·596 20·277	2·547 18·027	2·504 16·257
0·560	2·803 33·293	2·727 27·277	2·663 23·197	2·608 20·240	2·560 17·994	2·517 16·227
0·565	2·816 33·226	2·739 27·223	2·676 23·151	2·621 20·200	2·573 17·958	2·529 16·195
0·570	2·828 33·155	2·752 27·164	2·689 23·101	2·634 20·156	2·585 17·919	2·542 16·160
0·575	2·841 33·078	2·765 27·101	2·701 23·047	2·646 20·109	2·598 17·878	2·555 16·122
0·580	2·854 32·995	2·778 27·033	2·714 22·990	2·659 20·059	2·611 17·833	2·568 16·082
0·585	2·867 32·907	2·791 26·961	2·727 22·929	2·672 20·006	2·624 17·786	2·580 16·039
0·590	2·880 32·814	2·803 26·885	2·740 22·864	2·685 19·949	2·636 17·735	2·593 15·994
0·595	2·892 32·715	2·816 26·804	2·753 22·795	2·698 19·889	2·649 17·682	2·606 15·945
0·600	2·905 32·611	2·829 26·718	2·765 22·722	2·711 19·826	2·662 17·625	2·619 15·895
0·605	2·918 32·501	2·842 26·629	2·778 22·646	2·724 19·759	2·675 17·566	2·632 15·841
0·610	2·931 32·386	2·855 26·534	2·791 22·566	2·737 19·689	2·688 17·504	2·645 15·785
0·615	2·944 32·265	2·868 26·435	2·805 22·482	2·750 19·616	2·701 17·439	2·658 15·726
0·620	2·958 32·139	2·881 26·332	2·818 22·394	2·763 19·539	2·714 17·371	2·671 15·665
0·625	2·971 32·008	2·894 26·224	2·831 22·302	2·776 19·459	2·728 17·299	2·684 15·601
0·630	2·984 31·870	2·908 26·112	2·844 22·206	2·789 19·375	2·741 17·225	2·697 15·534
0·635	2·997 31·727	2·921 25·995	2·857 22·107	2·802 19·289	2·754 17·148	2·711 15·464
0·640	3·011 31·579	2·934 25·873	2·871 22·003	2·816 19·198	2·767 17·068	2·724 15·392
0·645	3·024 31·425	2·948 25·747	2·884 21·896	2·829 19·105	2·781 16·984	2·737 15·317
0·650	3·037 31·265	2·961 25·616	2·897 21·784	2·843 19·008	2·794 16·898	2·751 15·239
Cutoff	2·652	2·576	2·512	2·457	2·409	2·366

Hit Rate	0·0040	0·0050	0·0060	0·0070	0·0080	0·0090
0·655	3·051 31·100	2·975 25·480	2·911 21·669	2·856 18·907	2·808 16·809	2·764 15·158
0·660	3·065 30·928	2·988 25·340	2·925 21·550	2·870 18·803	2·821 16·716	2·778 15·075
0·665	3·078 30·751	3·002 25·195	2·938 21·427	2·883 18·695	2·835 16·620	2·792 14·988
0·670	3·092 30·569	3·016 25·045	2·952 21·299	2·897 18·584	2·849 16·522	2·806 14·899
0·675	3·106 30·380	3·030 24·891	2·966 21·168	2·911 18·470	2·863 16·420	2·819 14·807
0·680	3·120 30·186	3·044 24·731	2·980 21·032	2·925 18·351	2·877 16·315	2·833 14·713
0·685	3·134 29·985	3·058 24·567	2·994 20·893	2·939 18·230	2·891 16·206	2·847 14·615
0·690	3·148 29·779	3·072 24·398	3·008 20·749	2·953 18·104	2·905 16·095	2·861 14·514
0·695	3·162 29·567	3·086 24·224	3·022 20·601	2·967 17·975	2·919 15·980	2·876 14·411
0·700	3·176 29·348	3·100 24·046	3·037 20·449	2·982 17·842	2·933 15·862	2·890 14·305
0·710	3·205 28·894	3·129 23·673	3·066 20·132	3·011 17·566	2·962 15·616	2·919 14·083
0·720	3·235 28·414	3·159 23·280	3·095 19·798	3·040 17·274	2·992 15·357	2·948 13·849
0·730	3·265 27·910	3·189 22·867	3·125 19·446	3·070 16·968	3·022 15·084	2·978 13·603
0·740	3·295 27·379	3·219 22·432	3·155 19·077	3·101 16·645	3·052 14·798	3·009 13·345
0·750	3·327 26·823	3·250 21·977	3·187 18·690	3·132 16·307	3·083 14·497	3·040 13·074
0·760	3·358 26·241	3·282 21·499	3·218 18·284	3·164 15·953	3·115 14·182	3·072 12·790
0·770	3·391 25·631	3·315 21·000	3·251 17·859	3·196 15·582	3·148 13·853	3·104 12·492
0·780	3·424 24·993	3·348 20·477	3·284 17·414	3·229 15·194	3·181 13·508	3·138 12·182
0·790	3·458 24·327	3·382 19·931	3·319 16·950	3·264 14·789	3·215 13·148	3·172 11·857
0·800	3·494 23·631	3·417 19·361	3·354 16·466	3·299 14·367	3·251 12·772	3·207 11·518
Cutoff	2·652	2·576	2·512	2·457	2·409	2·366

FALSE POSITIVE RATE

Hit Rate	0·0040	0·0050	0·0060	0·0070	0·0080	0·0090
0·810	3·530 22·906	3·454 18·767	3·390 15·960	3·335 13·925	3·287 12·380	3·244 11·164
0·820	3·567 22·149	3·491 18·147	3·428 15·433	3·373 13·465	3·324 11·971	3·281 10·795
0·830	3·606 21·360	3·530 17·501	3·466 14·883	3·411 12·986	3·363 11·545	3·320 10·411
0·840	3·647 20·538	3·570 16·827	3·507 14·310	3·452 12·486	3·403 11·100	3·360 10·010
0·850	3·689 19·681	3·612 16·125	3·549 13·713	3·494 11·965	3·445 10·637	3·402 9·592
0·860	3·732 18·788	3·656 15·393	3·592 13·091	3·538 11·422	3·489 10·154	3·446 9·157
0·870	3·778 17·856	3·702 14·630	3·639 12·442	3·584 10·856	3·535 9·651	3·492 8·703
0·880	3·827 16·885	3·751 13·834	3·687 11·765	3·632 10·265	3·584 9·126	3·541 8·230
0·890	3·879 15·872	3·802 13·004	3·739 11·059	3·684 9·649	3·635 8·578	3·592 7·736
0·900	3·934 14·814	3·857 12·137	3·794 10·322	3·739 9·006	3·690 8·006	3·647 7·220
0·910	3·993 13·707	3·917 11·231	3·853 9·551	3·798 8·333	3·750 7·408	3·706 6·681
0·920	4·057 12·549	3·981 10·281	3·917 8·744	3·862 7·629	3·814 6·782	3·771 6·116
0·930	4·128 11·333	4·052 9·286	3·988 7·897	3·933 6·890	3·885 6·125	3·841 5·524
0·940	4·207 10·055	4·131 8·238	4·067 7·006	4·012 6·113	3·964 5·435	3·920 4·901
0·950	4·297 8·706	4·221 7·133	4·157 6·066	4·102 5·293	4·054 4·705	4·010 4·243
0·960	4·403 7·274	4·327 5·960	4·263 5·068	4·208 4·422	4·160 3·931	4·116 3·545
0·970	4·533 5·743	4·457 4·706	4·393 4·002	4·338 3·492	4·290 3·104	4·246 2·799
0·980	4·706 4·087	4·630 3·348	4·566 2·848	4·511 2·485	4·463 2·209	4·419 1·992
0·990	4·978 2·250	4·902 1·843	4·838 1·568	4·784 1·368	4·735 1·216	4·692 1·097
0·999	5·742 0·284	5·666 0·233	5·602 0·198	5·547 0·173	5·499 0·154	5·456 0·139
Cutoff	2·652	2·576	2·512	2·457	2·409	2·366

Hit Rate	0·0100	0·0120	0·0140	0·0160	0·0180	0·0200
0·010	0·000	−0·069	−0·129	−0·182	−0·229	−0·273
	1·000	0·853	0·747	0·666	0·602	0·550
0·020	0·273	0·203	0·144	0·091	0·043	0·000
	1·817	1·550	1·357	1·210	1·094	1·000
0·030	0·446	0·376	0·316	0·264	0·216	0·173
	2·553	2·178	1·907	1·700	1·537	1·405
0·040	0·576	0·506	0·447	0·394	0·346	0·303
	3·233	2·759	2·415	2·153	1·947	1·780
0·050	0·681	0·612	0·552	0·500	0·452	0·409
	3·870	3·302	2·890	2·577	2·330	2·130
0·060	0·772	0·702	0·643	0·590	0·542	0·499
	4·470	3·814	3·338	2·976	2·691	2·460
0·070	0·851	0·781	0·721	0·669	0·621	0·578
	5·038	4·299	3·762	3·354	3·033	2·773
0·080	0·921	0·852	0·792	0·739	0·692	0·649
	5·578	4·760	4·166	3·714	3·358	3·070
0·090	0·986	0·916	0·857	0·804	0·756	0·713
	6·093	5·199	4·550	4·057	3·668	3·354
0·100	1·045	0·976	0·916	0·863	0·815	0·772
	6·585	5·619	4·918	4·384	3·964	3·625
0·110	1·100	1·031	0·971	0·918	0·870	0·827
	7·055	6·020	5·269	4·698	4·248	3·884
0·120	1·151	1·082	1·022	0·969	0·922	0·879
	7·506	6·405	5·605	4·998	4·519	4·132
0·130	1·200	1·131	1·071	1·018	0·971	0·927
	7·937	6·773	5·928	5·285	4·779	4·369
0·140	1·246	1·177	1·117	1·064	1·017	0·973
	8·351	7·126	6·237	5·561	5·028	4·597
0·150	1·290	1·221	1·161	1·108	1·060	1·017
	8·748	7·465	6·533	5·825	5·267	4·816
0·160	1·332	1·263	1·203	1·150	1·102	1·059
	9·129	7·790	6·818	6·079	5·496	5·025
0·170	1·372	1·303	1·243	1·190	1·143	1·100
	9·495	8·102	7·091	6·322	5·716	5·226
0·180	1·411	1·342	1·282	1·229	1·182	1·138
	9·845	8·401	7·353	6·555	5·928	5·419
0·190	1·448	1·379	1·319	1·267	1·219	1·176
	10·182	8·688	7·604	6·779	6·130	5·605
0·200	1·485	1·416	1·356	1·303	1·255	1·212
	10·504	8·963	7·845	6·994	6·324	5·782
Cutoff	2·326	2·257	2·197	2·144	2·097	2·054

Hit Rate	0·0100	0·0120	0·0140	0·0160	0·0180	0·0200
0·210	1·520	1·451	1·391	1·338	1·291	1·247
	10·813	9·227	8·076	7·200	6·510	5·952
0·220	1·554	1·485	1·425	1·372	1·325	1·282
	11·110	9·480	8·297	7·397	6·689	6·115
0·230	1·588	1·518	1·458	1·406	1·358	1·315
	11·393	9·722	8·509	7·586	6·859	6·271
0·240	1·620	1·551	1·491	1·438	1·391	1·347
	11·664	9·953	8·711	7·766	7·023	6·421
0·250	1·652	1·583	1·523	1·470	1·422	1·379
	11·923	10·174	8·905	7·939	7·178	6·563
0·260	1·683	1·614	1·554	1·501	1·454	1·410
	12·170	10·385	9·089	8·104	7·327	6·699
0·270	1·714	1·644	1·584	1·532	1·484	1·441
	12·406	10·586	9·265	8·260	7·469	6·829
0·280	1·744	1·674	1·614	1·562	1·514	1·471
	12·630	10·778	9·433	8·410	7·604	6·952
0·290	1·773	1·704	1·644	1·591	1·544	1·500
	12·843	10·959	9·592	8·552	7·733	7·070
0·300	1·802	1·733	1·673	1·620	1·573	1·529
	13·046	11·132	9·743	8·686	7·854	7·181
0·305	1·816	1·747	1·687	1·634	1·587	1·544
	13·143	11·215	9·815	8·751	7·913	7·234
0·310	1·830	1·761	1·701	1·649	1·601	1·558
	13·237	11·295	9·886	8·814	7·969	7·286
0·315	1·845	1·775	1·716	1·663	1·615	1·572
	13·329	11·373	9·954	8·875	8·025	7·337
0·320	1·859	1·789	1·730	1·677	1·629	1·586
	13·418	11·449	10·021	8·934	8·078	7·386
0·325	1·873	1·803	1·744	1·691	1·643	1·600
	13·504	11·523	10·085	8·992	8·130	7·433
0·330	1·886	1·817	1·757	1·704	1·657	1·614
	13·588	11·595	10·148	9·047	8·181	7·480
0·335	1·900	1·831	1·771	1·718	1·671	1·628
	13·669	11·664	10·209	9·102	8·230	7·524
0·340	1·914	1·845	1·785	1·732	1·684	1·641
	13·748	11·731	10·267	9·154	8·277	7·568
0·345	1·927	1·858	1·798	1·746	1·698	1·655
	13·824	11·796	10·324	9·205	8·323	7·610
0·350	1·941	1·872	1·812	1·759	1·712	1·668
	13·898	11·859	10·379	9·254	8·367	7·650
Cutoff	2·326	2·257	2·197	2·144	2·097	2·054

Hit Rate	0·0100	0·0120	0·0140	0·0160	0·0180	0·0200
0·355	1·954 13·969	1·885 11·920	1·825 10·432	1·773 9·301	1·725 8·410	1·682 7·689
0·360	1·968 14·037	1·899 11·978	1·839 10·483	1·786 9·347	1·738 8·451	1·695 7·727
0·365	1·981 14·103	1·912 12·034	1·852 10·533	1·799 9·390	1·752 8·491	1·709 7·763
0·370	1·994 14·167	1·925 12·088	1·865 10·580	1·813 9·433	1·765 8·529	1·722 7·798
0·375	2·008 14·228	1·938 12·141	1·879 10·626	1·826 9·473	1·778 8·566	1·735 7·832
0·380	2·021 14·286	1·952 12·190	1·892 10·669	1·839 9·512	1·791 8·601	1·748 7·864
0·385	2·034 14·342	1·965 12·238	1·905 10·711	1·852 9·550	1·805 8·635	1·761 7·895
0·390	2·047 14·396	1·978 12·284	1·918 10·751	1·865 9·585	1·818 8·667	1·774 7·924
0·395	2·060 14·447	1·991 12·328	1·931 10·789	1·878 9·619	1·831 8·698	1·787 7·952
0·400	2·073 14·496	2·004 12·369	1·944 10·826	1·891 9·652	1·844 8·727	1·800 7·979
0·405	2·086 14·542	2·017 12·409	1·957 10·860	1·904 9·683	1·857 8·755	1·813 8·005
0·410	2·099 14·586	2·030 12·446	1·970 10·893	1·917 9·712	1·869 8·782	1·826 8·029
0·415	2·112 14·627	2·042 12·482	1·983 10·924	1·930 9·740	1·882 8·807	1·839 8·052
0·420	2·124 14·667	2·055 12·515	1·995 10·953	1·943 9·766	1·895 8·830	1·852 8·073
0·425	2·137 14·703	2·068 12·546	2·008 10·981	1·955 9·790	1·908 8·852	1·865 8·093
0·430	2·150 14·737	2·081 12·576	2·021 11·006	1·968 9·813	1·921 8·873	1·877 8·112
0·435	2·163 14·769	2·093 12·603	2·034 11·030	1·981 9·834	1·933 8·892	1·890 8·130
0·440	2·175 14·799	2·106 12·628	2·046 11·052	1·993 9·854	1·946 8·910	1·903 8·146
0·445	2·188 14·826	2·119 12·651	2·059 11·073	2·006 9·872	1·959 8·926	1·915 8·161
0·450	2·201 14·851	2·131 12·672	2·072 11·091	2·019 9·888	1·971 8·941	1·928 8·175
Cutoff	2·326	2·257	2·197	2·144	2·097	2·054

Hit Rate	0·0100	0·0120	0·0140	0·0160	0·0180	0·0200
0·455	2·213 14·873	2·144 12·691	2·084 11·108	2·031 9·903	1·984 8·955	1·941 8·187
0·460	2·226 14·893	2·157 12·709	2·097 11·123	2·044 9·917	1·996 8·967	1·953 8·198
0·465	2·239 14·911	2·169 12·724	2·109 11·136	2·057 9·928	2·009 8·977	1·966 8·208
0·470	2·251 14·926	2·182 12·737	2·122 11·147	2·069 9·938	2·022 8·986	1·978 8·216
0·475	2·264 14·939	2·194 12·748	2·135 11·157	2·082 9·947	2·034 8·994	1·991 8·223
0·480	2·276 14·950	2·207 12·757	2·147 11·165	2·094 9·954	2·047 9·001	2·004 8·229
0·485	2·289 14·958	2·220 12·764	2·160 11·171	2·107 9·960	2·059 9·006	2·016 8·234
0·490	2·301 14·964	2·232 12·769	2·172 11·175	2·119 9·964	2·072 9·009	2·029 8·237
0·495	2·314 14·967	2·245 12·772	2·185 11·178	2·132 9·966	2·084 9·011	2·041 8·239
0·500	2·326 14·968	2·257 12·773	2·197 11·179	2·144 9·967	2·097 9·012	2·054 8·240
0·505	2·339 14·967	2·270 12·772	2·210 11·178	2·157 9·966	2·109 9·011	2·066 8·239
0·510	2·351 14·964	2·282 12·769	2·222 11·175	2·169 9·964	2·122 9·009	2·079 8·237
0·515	2·364 14·958	2·295 12·764	2·235 11·171	2·182 9·960	2·135 9·006	2·091 8·234
0·520	2·377 14·950	2·307 12·757	2·247 11·165	2·195 9·954	2·147 9·001	2·104 8·229
0·525	2·389 14·939	2·320 12·748	2·260 11·157	2·207 9·947	2·160 8·994	2·116 8·223
0·530	2·402 14·926	2·332 12·737	2·273 11·147	2·220 9·938	2·172 8·986	2·129 8·216
0·535	2·414 14·911	2·345 12·724	2·285 11·136	2·232 9·928	2·185 8·977	2·142 8·208
0·540	2·427 14·893	2·358 12·709	2·298 11·123	2·245 9·917	2·197 8·967	2·154 8·198
0·545	2·439 14·873	2·370 12·691	2·310 11·108	2·257 9·903	2·210 8·955	2·167 8·187
0·550	2·452 14·851	2·383 12·672	2·323 11·091	2·270 9·888	2·223 8·941	2·179 8·175
Cutoff	2·326	2·257	2·197	2·144	2·097	2·054

Hit Rate	0·0100	0·0120	0·0140	0·0160	0·0180	0·0200
0·555	2·465	2·395	2·336	2·283	2·235	2·192
	14·826	12·651	11·073	9·872	8·926	8·161
0·560	2·477	2·408	2·348	2·295	2·248	2·205
	14·799	12·628	11·052	9·854	8·910	8·146
0·565	2·490	2·421	2·361	2·308	2·261	2·217
	14·769	12·603	11·030	9·834	8·892	8·130
0·570	2·503	2·434	2·374	2·321	2·273	2·230
	14·737	12·576	11·006	9·813	8·873	8·112
0·575	2·515	2·446	2·386	2·334	2·286	2·243
	14·703	12·546	10·981	9·790	8·852	8·093
0·580	2·528	2·459	2·399	2·346	2·299	2·256
	14·667	12·515	10·953	9·766	8·830	8·073
0·585	2·541	2·472	2·412	2·359	2·312	2·268
	14·627	12·482	10·924	9·740	8·807	8·052
0·590	2·554	2·485	2·425	2·372	2·324	2·281
	14·586	12·446	10·893	9·712	8·782	8·029
0·595	2·567	2·498	2·438	2·385	2·337	2·294
	14·542	12·409	10·860	9·683	8·755	8·005
0·600	2·580	2·510	2·451	2·398	2·350	2·307
	14·496	12·369	10·826	9·652	8·727	7·979
0·605	2·593	2·523	2·464	2·411	2·363	2·320
	14·447	12·328	10·789	9·619	8·698	7·952
0·610	2·606	2·536	2·477	2·424	2·376	2·333
	14·396	12·284	10·751	9·585	8·667	7·924
0·615	2·619	2·550	2·490	2·437	2·389	2·346
	14·342	12·238	10·711	9·550	8·635	7·895
0·620	2·632	2·563	2·503	2·450	2·402	2·359
	14·286	12·190	10·669	9·512	8·601	7·864
0·625	2·645	2·576	2·516	2 463	2·416	2·372
	14·228	12·141	10·626	9·473	8·566	7·832
0·630	2·658	2·589	2·529	2·476	2·429	2·386
	14·167	12·088	10·580	9·433	8·529	7·798
0·635	2·671	2·602	2·542	2·490	2·442	2·399
	14·103	12·034	10·533	9·390	8·491	7·763
0·640	2·685	2·616	2·556	2·503	2·455	2·412
	14·037	11·978	10·483	9·347	8·451	7·727
0·645	2·698	2·629	2·569	2·516	2·469	2·426
	13·969	11·920	10·432	9·301	8·410	7·689
0·650	2·712	2·642	2·583	2·530	2·482	2·439
	13·898	11·859	10·379	9·254	8·367	7·650
Cutoff	2·326	2·257	2·197	2·144	2·097	2·054

FALSE POSITIVE RATE

Hit Rate	0·0100	0·0120	0·0140	0·0160	0·0180	0·0200
0·655	2·725 13·824	2·656 11·796	2·596 10·324	2·543 9·205	2·496 8·323	2·453 7·610
0·660	2·739 13·748	2·670 11·731	2·610 10·267	2·557 9·154	2·509 8·277	2·466 7·568
0·665	2·752 13·669	2·683 11·664	2·623 10·209	2·571 9·102	2·523 8·230	2·480 7·524
0·670	2·766 13·588	2·697 11·595	2·637 10·148	2·584 9·047	2·537 8·181	2·494 7·480
0·675	2·780 13·504	2·711 11·523	2·651 10·085	2·598 8·992	2·551 8·130	2·508 7·433
0·680	2·794 13·418	2·725 11·449	2·665 10·021	2·612 8·934	2·565 8·078	2·521 7·386
0·685	2·808 13·329	2·739 11·373	2·679 9·954	2·626 8·875	2·579 8·025	2·535 7·337
0·690	2·822 13·237	2·753 11·295	2·693 9·886	2·640 8·814	2·593 7·969	2·550 7·286
0·695	2·836 13·143	2·767 11·215	2·707 9·815	2·654 8·751	2·607 7·913	2·564 7·234
0·700	2·851 13·046	2·782 11·132	2·722 9·743	2·669 8·686	2·621 7·854	2·578 7·181
0·710	2·880 12·843	2·811 10·959	2·751 9·592	2·698 8·552	2·650 7·733	2·607 7·070
0·720	2·909 12·630	2·840 10·778	2·780 9·433	2·727 8·410	2·680 7·604	2·637 6·952
0·730	2·939 12·406	2·870 10·586	2·810 9·265	2·757 8·260	2·710 7·469	2·667 6·829
0·740	2·970 12·170	2·900 10·385	2·841 9·089	2·788 8·104	2·740 7·327	2·697 6·699
0·750	3·001 11·923	2·932 10·174	2·872 8·905	2·819 7·939	2·771 7·178	2·728 6·563
0·760	3·033 11·664	2·963 9·953	2·904 8·711	2·851 7·766	2·803 7·023	2·760 6·421
0·770	3·065 11·393	2·996 9·722	2·936 8·509	2·883 7·586	2·836 6·859	2·793 6·271
0·780	3·099 11·110	3·029 9·480	2·969 8·297	2·917 7·397	2·869 6·689	2·826 6·115
0·790	3·133 10·813	3·064 9·227	3·004 8·076	2·951 7·200	2·903 6·510	2·860 5·952
0·800	3·168 10·504	3·099 8·963	3·039 7·845	2·986 6·994	2·939 6·324	2·895 5·782
Cutoff	2·326	2·257	2·197	2·144	2·097	2·054

Hit Rate	0·0100	0·0120	0·0140	0·0160	0·0180	0·0200
0·810	3·204 10·182	3·135 8·688	3·075 7·604	3·022 6·779	2·975 6·130	2·932 5·605
0·820	3·242 9·845	3·172 8·401	3·113 7·353	3·060 6·555	3·012 5·928	2·969 5·419
0·830	3·281 9·495	3·211 8·102	3·151 7·091	3·099 6·322	3·051 5·716	3·008 5·226
0·840	3·321 9·129	3·252 7·790	3·192 6·818	3·139 6·079	3·091 5·496	3·048 5·025
0·850	3·363 8·748	3·294 7·465	3·234 6·533	3·181 5·825	3·133 5·267	3·090 4·816
0·860	3·407 8·351	3·337 7·126	3·278 6·237	3·225 5·561	3·177 5·028	3·134 4·597
0·870	3·453 7·937	3·384 6·773	3·324 5·928	3·271 5·285	3·223 4·779	3·180 4·369
0·880	3·501 7·506	3·432 6·405	3·372 5·605	3·319 4·998	3·272 4·519	3·229 4·132
0·890	3·553 7·055	3·484 6·020	3·424 5·269	3·371 4·698	3·323 4·248	3·280 3·884
0·900	3·608 6·585	3·539 5·619	3·479 4·918	3·426 4·384	3·378 3·964	3·335 3·625
0·910	3·667 6·093	3·598 5·199	3·538 4·550	3·485 4·057	3·438 3·668	3·395 3·354
0·920	3·731 5·578	3·662 4·760	3·602 4·166	3·549 3·714	3·502 3·358	3·459 3·070
0·930	3·802 5·038	3·733 4·299	3·673 3·762	3·620 3·354	3·573 3·033	3·530 2·773
0·940	3·881 4·470	3·812 3·814	3·752 3·338	3·699 2·976	3·652 2·691	3·609 2·460
0·950	3·971 3·870	3·902 3·302	3·842 2·890	3·789 2·577	3·742 2·330	3·699 2·130
0·960	4·077 3·233	4·008 2·759	3·948 2·415	3·895 2·153	3·848 1·947	3·804 1·780
0·970	4·207 2·553	4·138 2·178	4·078 1·907	4·025 1·700	3·978 1·537	3·935 1·405
0·980	4·380 1·817	4·311 1·550	4·251 1·357	4·198 1·210	4·151 1·094	4·107 1·000
0·990	4·653 1·000	4·583 0·853	4·524 0·747	4·471 0·666	4·423 0·602	4·380 0·550
0·999	5·417 0·126	5·347 0·108	5·288 0·094	5·235 0·084	5·187 0·076	5·144 0·070
Cutoff	2·326	2·257	2·197	2·144	2·097	2·054

Hit Rate	0·0220	0·0240	0·0260	0·0280	0·0300	0·0320
0·010	−0·312 0·508	−0·349 0·472	−0·383 0·441	−0·415 0·415	−0·446 0·392	−0·474 0·371
0·020	−0·040 0·923	−0·076 0·857	−0·111 0·802	−0·143 0·754	−0·173 0·712	−0·202 0·675
0·030	0·133 1·296	0·097 1·205	0·062 1·127	0·030 1·059	0·000 1·000	−0·029 0·948
0·040	0·263 1·642	0·227 1·526	0·192 1·427	0·160 1·341	0·130 1·266	0·101 1·201
0·050	0·369 1·965	0·333 1·826	0·298 1·708	0·266 1·605	0·236 1·516	0·207 1·437
0·060	0·459 2·270	0·423 2·109	0·388 1·972	0·356 1·854	0·326 1·751	0·297 1·660
0·070	0·538 2·558	0·502 2·377	0·467 2·223	0·435 2·090	0·405 1·973	0·376 1·871
0·080	0·609 2·833	0·572 2·632	0·538 2·462	0·506 2·314	0·476 2·185	0·447 2·071
0·090	0·673 3·094	0·637 2·875	0·602 2·689	0·570 2·527	0·540 2·387	0·511 2·263
0·100	0·733 3·344	0·696 3·107	0·662 2·906	0·629 2·731	0·599 2·579	0·571 2·445
0·110	0·788 3·583	0·751 3·329	0·717 3·113	0·685 2·927	0·654 2·764	0·626 2·620
0·120	0·839 3·811	0·802 3·542	0·768 3·312	0·736 3·113	0·706 2·940	0·677 2·787
0·130	0·888 4·031	0·851 3·746	0·817 3·503	0·785 3·293	0·754 3·109	0·726 2·947
0·140	0·934 4·241	0·897 3·941	0·863 3·685	0·831 3·464	0·800 3·271	0·772 3·101
0·150	0·978 4·442	0·941 4·128	0·907 3·860	0·875 3·629	0·844 3·427	0·816 3·248
0·160	1·020 4·636	0·983 4·308	0·949 4·029	0·917 3·787	0·886 3·576	0·858 3·390
0·170	1·060 4·821	1·023 4·481	0·989 4·190	0·957 3·939	0·927 3·719	0·898 3·526
0·180	1·099 4·999	1·062 4·646	1·028 4·345	0·996 4·084	0·965 3·856	0·937 3·656
0·190	1·136 5·170	1·099 4·805	1·065 4·493	1·033 4·224	1·003 3·988	0·974 3·781
0·200	1·172 5·334	1·136 4·957	1·102 4·635	1·069 4·357	1·039 4·115	1·011 3·901
Cutoff	2·014	1·977	1·943	1·911	1·881	1·852

Hit Rate	0·0220	0·0240	0·0260	0·0280	0·0300	0·0320
0·210	1·208 5·491	1·171 5·103	1·137 4·772	1·105 4·486	1·074 4·236	1·046 4·015
0·220	1·242 5·641	1·205 5·243	1·171 4·902	1·139 4·608	1·109 4·352	1·080 4·125
0·230	1·275 5·785	1·239 5·377	1·204 5·028	1·172 4·726	1·142 4·463	1·113 4·231
0·240	1·308 5·923	1·271 5·504	1·237 5·147	1·205 4·838	1·174 4·569	1·146 4·331
0·250	1·340 6·055	1·303 5·627	1·269 5·262	1·237 4·946	1·206 4·670	1·178 4·427
0·260	1·371 6·180	1·334 5·743	1·300 5·371	1·268 5·048	1·237 4·767	1·209 4·519
0·270	1·401 6·300	1·365 5·855	1·330 5·475	1·298 5·146	1·268 4·859	1·239 4·607
0·280	1·431 6·414	1·395 5·960	1·360 5·574	1·328 5·239	1·298 4·947	1·269 4·690
0·290	1·461 6·522	1·424 6·061	1·390 5·668	1·358 5·328	1·327 5·031	1·299 4·769
0·300	1·490 6·625	1·453 6·156	1·419 5·757	1·387 5·412	1·356 5·110	1·328 4·844
0·305	1·504 6·674	1·467 6·202	1·433 5·800	1·401 5·452	1·371 5·148	1·342 4·880
0·310	1·518 6·722	1·482 6·247	1·447 5·841	1·415 5·491	1·385 5·185	1·356 4·915
0·315	1·532 6·768	1·496 6·290	1·461 5·882	1·429 5·529	1·399 5·221	1·370 4·949
0·320	1·546 6·814	1·510 6·332	1·475 5·921	1·443 5·566	1·413 5·256	1·384 4·982
0·325	1·560 6·857	1·524 6·373	1·489 5·959	1·457 5·602	1·427 5·290	1·398 5·015
0·330	1·574 6·900	1·537 6·412	1·503 5·996	1·471 5·637	1·441 5·322	1·412 5·046
0·335	1·588 6·941	1·551 6·451	1·517 6·032	1·485 5·670	1·455 5·354	1·426 5·076
0·340	1·602 6·981	1·565 6·488	1·531 6·067	1·499 5·703	1·468 5·385	1·440 5·105
0·345	1·615 7·020	1·579 6·524	1·544 6·100	1·512 5·734	1·482 5·415	1·453 5·133
0·350	1·629 7·057	1·592 6·558	1·558 6·133	1·526 5·765	1·495 5·444	1·467 5·161
Cutoff	2·014	1·977	1·943	1·911	1·881	1·852

FALSE POSITIVE RATE

Hit Rate	0·0220	0·0240	0·0260	0·0280	0·0300	0·0320
0·355	1·642 7·093	1·606 6·592	1·571 6·164	1·539 5·794	1·509 5·472	1·480 5·187
0·360	1·656 7·128	1·619 6·624	1·585 6·194	1·553 5·823	1·522 5·498	1·494 5·212
0·365	1·669 7·162	1·632 6·655	1·598 6·223	1·566 5·850	1·536 5·524	1·507 5·237
0·370	1·682 7·194	1·646 6·685	1·611 6·251	1·579 5·877	1·549 5·549	1·520 5·260
0·375	1·695 7·225	1·659 6·714	1·624 6·278	1·592 5·902	1·562 5·573	1·534 5·283
0·380	1·709 7·254	1·672 6·742	1·638 6·304	1·606 5·926	1·575 5·596	1·547 5·305
0·385	1·722 7·283	1·685 6·768	1·651 6·329	1·619 5·949	1·588 5·618	1·560 5·326
0·390	1·735 7·310	1·698 6·794	1·664 6·353	1·632 5·972	1·601 5·639	1·573 5·346
0·395	1·748 7·336	1·711 6·818	1·677 6·375	1·645 5·993	1·614 5·659	1·586 5·365
0·400	1·761 7·361	1·724 6·841	1·690 6·397	1·658 6·013	1·627 5·678	1·599 5·383
0·405	1·774 7·384	1·737 6·863	1·703 6·417	1·671 6·032	1·640 5·696	1·612 5·400
0·410	1·787 7·407	1·750 6·883	1·716 6·437	1·683 6·050	1·653 5·713	1·625 5·416
0·415	1·799 7·428	1·763 6·903	1·728 6·455	1·696 6·068	1·666 5·730	1·637 5·432
0·420	1·812 7·448	1·775 6·921	1·741 6·472	1·709 6·084	1·679 5·745	1·650 5·446
0·425	1·825 7·466	1·788 6·939	1·754 6·488	1·722 6·099	1·692 5·759	1·663 5·460
0·430	1·838 7·484	1·801 6·955	1·767 6·503	1·735 6·113	1·704 5·773	1·676 5·472
0·435	1·850 7·500	1·814 6·970	1·779 6·518	1·747 6·127	1·717 5·785	1·689 5·484
0·440	1·863 7·515	1·826 6·984	1·792 6·531	1·760 6·139	1·730 5·797	1·701 5·495
0·445	1·876 7·529	1·839 6·997	1·805 6·543	1·773 6·150	1·742 5·807	1·714 5·505
0·450	1·888 7·541	1·852 7·008	1·817 6·553	1·785 6·160	1·755 5·817	1·727 5·515
Cutoff	2·014	1·977	1·943	1·911	1·881	1·852

Hit Rate	0·0220	0·0240	0·0260	0·0280	0·0300	0·0320
0·455	1·901	1·864	1·830	1·798	1·768	1·739
	7·553	7·019	6·563	6·170	5·826	5·523
0·460	1·914	1·877	1·843	1·811	1·780	1·752
	7·563	7·028	6·572	6·178	5·834	5·530
0·465	1·926	1·890	1·855	1·823	1·793	1·764
	7·572	7·037	6·580	6·185	5·841	5·537
0·470	1·939	1·902	1·868	1·836	1·806	1·777
	7·580	7·044	6·587	6·192	5·847	5·543
0·475	1·951	1·915	1·880	1·848	1·818	1·789
	7·586	7·050	6·592	6·197	5·852	5·547
0·480	1·964	1·927	1·893	1·861	1·831	1·802
	7·591	7·055	6·597	6·201	5·856	5·551
0·485	1·976	1·940	1·906	1·873	1·843	1·815
	7·596	7·059	6·601	6·205	5·859	5·554
0·490	1·989	1·952	1·918	1·886	1·856	1·827
	7·599	7·062	6·603	6·207	5·861	5·557
0·495	2·002	1·965	1·931	1·899	1·868	1·840
	7·600	7·063	6·605	6·209	5·863	5·558
0·500	2·014	1·977	1·943	1·911	1·881	1·852
	7·601	7·064	6·605	6·209	5·863	5·558
0·505	2·027	1·990	1·956	1·924	1·893	1·865
	7·600	7·063	6·605	6·209	5·863	5·558
0·510	2·039	2·002	1·968	1·936	1·906	1·877
	7·599	7·062	6·603	6·207	5·861	5·557
0·515	2·052	2·015	1·981	1·949	1·918	1·890
	7·596	7·059	6·601	6·205	5·859	5·554
0·520	2·064	2·028	1·993	1·961	1·931	1·902
	7·591	7·055	6·597	6·201	5·856	5·551
0·525	2·077	2·040	2·006	1·974	1·944	1·915
	7·586	7·050	6·592	6·197	5·852	5·547
0·530	2·089	2·053	2·018	1·986	1·956	1·927
	7·580	7·044	6·587	6·192	5·847	5·543
0·535	2·102	2·065	2·031	1·999	1·969	1·940
	7·572	7·037	6·580	6·185	5·841	5·537
0·540	2·115	2·078	2·044	2·011	1·981	1·953
	7·563	7·028	6·572	6·178	5·834	5·530
0·545	2·127	2·090	2·056	2·024	1·994	1·965
	7·553	7·019	6·563	6·170	5·826	5·523
0·550	2·140	2·103	2·069	2·037	2·006	1·978
	7·541	7·008	6·553	6·160	5·817	5·515
Cutoff	2·014	1·977	1·943	1·911	1·881	1·852

Hit Rate	0·0220	0·0240	0·0260	0·0280	0·0300	0·0320
0·555	2·152 7·529	2·116 6·997	2·081 6·543	2·049 6·150	2·019 5·807	1·990 5·505
0·560	2·165 7·515	2·128 6·984	2·094 6·531	2·062 6·139	2·032 5·797	2·003 5·495
0·565	2·178 7·500	2·141 6·970	2·107 6·518	2·075 6·127	2·044 5·785	2·016 5·484
0·570	2·190 7·484	2·154 6·955	2·120 6·503	2·087 6·113	2·057 5·773	2·029 5·472
0·575	2·203 7·466	2·166 6·939	2·132 6·488	2·100 6·099	2·070 5·759	2·041 5·460
0·580	2·216 7·448	2·179 6·921	2·145 6·472	2·113 6·084	2·083 5·745	2·054 5·446
0·585	2·229 7·428	2·192 6·903	2·158 6·455	2·126 6·068	2·095 5·730	2·067 5·432
0·590	2·242 7·407	2·205 6·883	2·171 6·437	2·139 6·050	2·108 5·713	2·080 5·416
0·595	2·255 7·384	2·218 6·863	2·184 6·417	2·151 6·032	2·121 5·696	2·093 5·400
0·600	2·267 7·361	2·231 6·841	2·196 6·397	2·164 6·013	2·134 5·678	2·106 5·383
0·605	2·280 7·336	2·244 6·818	2·209 6·375	2·177 5·993	2·147 5·659	2·118 5·365
0·610	2·293 7·310	2·257 6·794	2·222 6·353	2·190 5·972	2·160 5·639	2·131 5·346
0·615	2·306 7·283	2·270 6·768	2·236 6·329	2·203 5·949	2·173 5·618	2·145 5·326
0·620	2·320 7·254	2·283 6·742	2·249 6·304	2·217 5·926	2·186 5·596	2·158 5·305
0·625	2·333 7·225	2·296 6·714	2·262 6·278	2·230 5·902	2·199 5·573	2·171 5·283
0·630	2·346 7·194	2·309 6·685	2·275 6·251	2·243 5·877	2·213 5·549	2·184 5·260
0·635	2·359 7·162	2·322 6·655	2·288 6·223	2·256 5·850	2·226 5·524	2·197 5·237
0·640	2·373 7·128	2·336 6·624	2·302 6·194	2·269 5·823	2·239 5·498	2·211 5·212
0·645	2·386 7·093	2·349 6·592	2·315 6·164	2·283 5·794	2·253 5·472	2·224 5·187
0·650	2·399 7·057	2·363 6·558	2·328 6·133	2·296 5·765	2·266 5·444	2·238 5·161
Cutoff	2·014	1·977	1·943	1·911	1·881	1·852

Hit Rate	0·0220	0·0240	0·0260	0·0280	0·0300	0·0320
0·655	2·413	2·376	2·342	2·310	2·280	2·251
	7·020	6·524	6·100	5·734	5·415	5·133
0·660	2·427	2·390	2·356	2·323	2·293	2·265
	6·981	6·488	6·067	5·703	5·385	5·105
0·665	2·440	2·404	2·369	2·337	2·307	2·278
	6·941	6·451	6·032	5·670	5·354	5·076
0·670	2·454	2·417	2·383	2·351	2·321	2·292
	6·900	6·412	5·996	5·637	5·322	5·046
0·675	2·468	2·431	2·397	2·365	2·335	2·306
	6·857	6·373	5·959	5·602	5·290	5·015
0·680	2·482	2·445	2·411	2·379	2·348	2·320
	6·814	6·332	5·921	5·566	5·256	4·982
0·685	2·496	2·459	2·425	2·393	2·363	2·334
	6·768	6·290	5·882	5·529	5·221	4·949
0·690	2·510	2·473	2·439	2·407	2·377	2·348
	6·722	6·247	5·841	5·491	5·185	4·915
0·695	2·524	2·487	2·453	2·421	2·391	2·362
	6·674	6·202	5·800	5·452	5·148	4·880
0·700	2·538	2·502	2·468	2·435	2·405	2·377
	6·625	6·156	5·757	5·412	5·110	4·844
0·710	2·567	2·531	2·497	2·464	2·434	2·406
	6·522	6·061	5·668	5·328	5·031	4·769
0·720	2·597	2·560	2·526	2·494	2·464	2·435
	6·414	5·960	5·574	5·239	4·947	4·690
0·730	2·627	2·590	2·556	2·524	2·494	2·465
	6·300	5·855	5·475	5·146	4·859	4·607
0·740	2·657	2·621	2·586	2·554	2·524	2·496
	6·180	5·743	5·371	5·048	4·767	4·519
0·750	2·689	2·652	2·618	2·586	2·555	2·527
	6·055	5·627	5·262	4·946	4·670	4·427
0·760	2·720	2·684	2·649	2·617	2·587	2·558
	5·923	5·504	5·147	4·838	4·569	4·331
0·770	2·753	2·716	2·682	2·650	2·620	2·591
	5·785	5·377	5·028	4·726	4·463	4·231
0·780	2·786	2·750	2·715	2·683	2·653	2·624
	5·641	5·243	4·902	4·608	4·352	4·125
0·790	2·821	2·784	2·750	2·717	2·687	2·659
	5·491	5·103	4·772	4·486	4·236	4·015
0·800	2·856	2·819	2·785	2·753	2·722	2·694
	5·334	4·957	4·635	4·357	4·115	3·901
Cutoff	2·014	1·977	1·943	1·911	1·881	1·852

Hit Rate	0·0220	0·0240	0·0260	0·0280	0·0300	0·0320
0·810	2·892	2·855	2·821	2·789	2·759	2·730
	5·170	4·805	4·493	4·224	3·988	3·781
0·820	2·929	2·893	2·858	2·826	2·796	2·768
	4·999	4·646	4·345	4·084	3·856	3·656
0·830	2·968	2·932	2·897	2·865	2·835	2·806
	4·821	4·481	4·190	3·939	3·719	3·526
0·840	3·009	2·972	2·938	2·905	2·875	2·847
	4·636	4·308	4·029	3·787	3·576	3·390
0·850	3·051	3·014	2·980	2·947	2·917	2·889
	4·442	4·128	3·860	3·629	3·427	3·248
0·860	3·094	3·058	3·023	2·991	2·961	2·932
	4·241	3·941	3·685	3·464	3·271	3·101
0·870	3·140	3·104	3·070	3·037	3·007	2·979
	4·031	3·746	3·503	3·293	3·109	2·947
0·880	3·189	3·152	3·118	3·086	3·056	3·027
	3·811	3·542	3·312	3·113	2·940	2·787
0·890	3·241	3·204	3·170	3·138	3·107	3·079
	3·583	3·329	3·113	2·927	2·764	2·620
0·900	3·296	3·259	3·225	3·193	3·162	3·134
	3·344	3·107	2·906	2·731	2·579	2·445
0·910	3·355	3·318	3·284	3·252	3·222	3·193
	3·094	2·875	2·689	2·527	2·387	2·263
0·920	3·419	3·382	3·348	3·316	3·286	3·257
	2·833	2·632	2·462	2·314	2·185	2·071
0·930	3·490	3·453	3·419	3·387	3·357	3·328
	2·558	2·377	2·223	2·090	1·973	1·871
0·940	3·569	3·532	3·498	3·466	3·436	3·407
	2·270	2·109	1·972	1·854	1·751	1·660
0·950	3·659	3·622	3·588	3·556	3·526	3·497
	1·965	1·826	1·708	1·605	1·516	1·437
0·960	3·765	3·728	3·694	3·662	3·631	3·603
	1·642	1·526	1·427	1·341	1·266	1·201
0·970	3·895	3·858	3·824	3·792	3·762	3·733
	1·296	1·205	1·127	1·059	1·000	0·948
0·980	4·068	4·031	3·997	3·965	3·935	3·906
	0·923	0·857	0·802	0·754	0·712	0·675
0·990	4·340	4·304	4·269	4·237	4·207	4·179
	0·508	0·472	0·441	0·415	0·392	0·371
0·999	5·104	5·068	5·033	5·001	4·971	4·942
	0·064	0·060	0·056	0·052	0·049	0·047
Cutoff	2·014	1·977	1·943	1·911	1·881	1·852

Hit Rate	0·0340	0·0360	0·0380	0·0400	0·0420	0·0440
0·010	−0·501	−0·527	−0·552	−0·576	−0·598	−0·620
	0·353	0·337	0·322	0·309	0·297	0·286
0·020	−0·229	−0·255	−0·279	−0·303	−0·326	−0·348
	0·642	0·612	0·586	0·562	0·540	0·520
0·030	−0·056	−0·082	−0·106	−0·130	−0·153	−0·175
	0·902	0·860	0·823	0·790	0·759	0·731
0·040	0·074	0·048	0·024	0·000	−0·023	−0·045
	1·142	1·090	1·043	1·000	0·961	0·926
0·050	0·180	0·154	0·130	0·106	0·083	0·061
	1·367	1·304	1·248	1·197	1·150	1·108
0·060	0·270	0·244	0·220	0·196	0·173	0·151
	1·579	1·506	1·441	1·382	1·329	1·280
0·070	0·349	0·323	0·299	0·275	0·252	0·230
	1·780	1·698	1·625	1·558	1·498	1·442
0·080	0·420	0·394	0·369	0·346	0·323	0·301
	1·970	1·880	1·799	1·725	1·658	1·597
0·090	0·484	0·458	0·434	0·410	0·387	0·365
	2·152	2·054	1·965	1·884	1·811	1·745
0·100	0·543	0·518	0·493	0·469	0·446	0·424
	2·326	2·219	2·123	2·037	1·958	1·885
0·110	0·598	0·573	0·548	0·524	0·501	0·480
	2·492	2·378	2·275	2·182	2·097	2·020
0·120	0·650	0·624	0·599	0·576	0·553	0·531
	2·651	2·530	2·420	2·321	2·231	2·149
0·130	0·699	0·673	0·648	0·624	0·602	0·580
	2·804	2·675	2·560	2·455	2·360	2·273
0·140	0·745	0·719	0·694	0·670	0·648	0·626
	2·950	2·815	2·693	2·583	2·483	2·391
0·150	0·789	0·763	0·738	0·714	0·692	0·670
	3·090	2·949	2·821	2·706	2·601	2·505
0·160	0·831	0·805	0·780	0·756	0·733	0·712
	3·225	3·077	2·944	2·824	2·714	2·614
0·170	0·871	0·845	0·820	0·797	0·774	0·752
	3·354	3·200	3·062	2·937	2·823	2·718
0·180	0·910	0·884	0·859	0·835	0·813	0·791
	3·478	3·318	3·175	3·045	2·927	2·819
0·190	0·947	0·921	0·896	0·873	0·850	0·828
	3·597	3·432	3·283	3·149	3·027	2·915
0·200	0·983	0·957	0·933	0·909	0·886	0·864
	3·710	3·540	3·387	3·249	3·123	3·008
Cutoff	1·825	1·799	1·774	1·751	1·728	1·706

Hit Rate	0·0340	0·0360	0·0380	0·0400	0·0420	0·0440
0·210	1·019	0·993	0·968	0·944	0·922	0·900
	3·820	3·645	3·487	3·344	3·215	3·096
0·220	1·053	1·027	1·002	0·978	0·956	0·934
	3·924	3·744	3·583	3·436	3·303	3·181
0·230	1·086	1·060	1·036	1·012	0·989	0·967
	4·024	3·840	3·674	3·524	3·387	3·262
0·240	1·119	1·093	1·068	1·044	1·022	1·000
	4·120	3·931	3·761	3·608	3·468	3·340
0·250	1·151	1·125	1·100	1·076	1·053	1·032
	4·212	4·019	3·845	3·688	3·545	3·414
0·260	1·182	1·156	1·131	1·107	1·085	1·063
	4·299	4·102	3·925	3·764	3·618	3·485
0·270	1·212	1·186	1·162	1·138	1·115	1·093
	4·382	4·181	4·001	3·837	3·688	3·552
0·280	1·242	1·216	1·192	1·168	1·145	1·123
	4·461	4·257	4·073	3·906	3·755	3·616
0·290	1·272	1·246	1·221	1·197	1·175	1·153
	4·537	4·329	4·142	3·972	3·818	3·677
0·300	1·301	1·275	1·250	1·226	1·204	1·182
	4·608	4·397	4·207	4·035	3·878	3·735
0·305	1·315	1·289	1·264	1·241	1·218	1·196
	4·642	4·430	4·238	4·065	3·907	3·763
0·310	1·329	1·303	1·279	1·255	1·232	1·210
	4·676	4·461	4·269	4·094	3·935	3·790
0·315	1·343	1·317	1·293	1·269	1·246	1·224
	4·708	4·492	4·298	4·122	3·962	3·816
0·320	1·357	1·331	1·307	1·283	1·260	1·238
	4·740	4·522	4·327	4·150	3·989	3·842
0·325	1·371	1·345	1·321	1·297	1·274	1·252
	4·770	4·552	4·355	4·177	4·015	3·866
0·330	1·385	1·359	1·334	1·311	1·288	1·266
	4·800	4·580	4·382	4·203	4·039	3·890
0·335	1·399	1·373	1·348	1·325	1·302	1·280
	4·828	4·607	4·408	4·228	4·064	3·914
0·340	1·413	1·387	1·362	1·338	1·315	1·294
	4·856	4·634	4·433	4·252	4·087	3·936
0·345	1·426	1·400	1·376	1·352	1·329	1·307
	4·883	4·659	4·458	4·276	4·110	3·958
0·350	1·440	1·414	1·389	1·365	1·343	1·321
	4·909	4·684	4·482	4·298	4·132	3·979
Cutoff	1·825	1·799	1·774	1·751	1·728	1·706

Hit Rate	0·0340	0·0360	0·0380	0·0400	0·0420	0·0440
0·355	1·453 4·934	1·427 4·708	1·403 4·505	1·379 4·320	1·356 4·153	1·334 3·999
0·360	1·467 4·958	1·441 4·731	1·416 4·527	1·392 4·341	1·369 4·173	1·348 4·019
0·365	1·480 4·982	1·454 4·753	1·429 4·548	1·406 4·362	1·383 4·193	1·361 4·038
0·370	1·493 5·004	1·467 4·775	1·443 4·568	1·419 4·381	1·396 4·211	1·374 4·056
0·375	1·506 5·026	1·480 4·795	1·456 4·588	1·432 4·400	1·409 4·230	1·387 4·074
0·380	1·520 5·046	1·494 4·815	1·469 4·607	1·445 4·418	1·422 4·247	1·401 4·090
0·385	1·533 5·066	1·507 4·834	1·482 4·625	1·458 4·436	1·436 4·264	1·414 4·106
0·390	1·546 5·085	1·520 4·852	1·495 4·642	1·471 4·452	1·449 4·280	1·427 4·122
0·395	1·559 5·103	1·533 4·869	1·508 4·659	1·484 4·468	1·462 4·295	1·440 4·136
0·400	1·572 5·120	1·546 4·886	1·521 4·675	1·497 4·483	1·475 4·309	1·453 4·150
0·405	1·585 5·137	1·559 4·901	1·534 4·689	1·510 4·498	1·488 4·323	1·466 4·164
0·410	1·597 5·152	1·572 4·916	1·547 4·704	1·523 4·511	1·500 4·336	1·478 4·176
0·415	1·610 5·167	1·584 4·930	1·560 4·717	1·536 4·524	1·513 4·348	1·491 4·188
0·420	1·623 5·181	1·597 4·943	1·572 4·730	1·549 4·536	1·526 4·360	1·504 4·199
0·425	1·636 5·194	1·610 4·956	1·585 4·741	1·562 4·547	1·539 4·371	1·517 4·210
0·430	1·649 5·206	1·623 4·967	1·598 4·752	1·574 4·558	1·552 4·381	1·530 4·220
0·435	1·661 5·217	1·635 4·978	1·611 4·763	1·587 4·568	1·564 4·391	1·542 4·229
0·440	1·674 5·227	1·648 4·988	1·623 4·772	1·600 4·577	1·577 4·399	1·555 4·237
0·445	1·687 5·237	1·661 4·997	1·636 4·781	1·612 4·585	1·590 4·408	1·568 4·245
0·450	1·699 5·246	1·673 5·005	1·649 4·789	1·625 4·593	1·602 4·415	1·580 4·252
Cutoff	1·825	1·799	1·774	1·751	1·728	1·706

Hit Rate	0·0340	0·0360	0·0380	0·0400	0·0420	0·0440
0·455	1·712 5·254	1·686 5·013	1·661 4·796	1·638 4·600	1·615 4·422	1·593 4·258
0·460	1·725 5·261	1·699 5·020	1·674 4·803	1·650 4·606	1·628 4·427	1·606 4·264
0·465	1·737 5·267	1·711 5·026	1·687 4·808	1·663 4·612	1·640 4·433	1·618 4·269
0·470	1·750 5·272	1·724 5·031	1·699 4·813	1·675 4·616	1·653 4·437	1·631 4·274
0·475	1·762 5·277	1·736 5·035	1·712 4·817	1·688 4·620	1·665 4·441	1·643 4·277
0·480	1·775 5·281	1·749 5·039	1·724 4·821	1·701 4·624	1·678 4·444	1·656 4·280
0·485	1·787 5·284	1·762 5·042	1·737 4·824	1·713 4·626	1·690 4·447	1·668 4·283
0·490	1·800 5·286	1·774 5·043	1·749 4·825	1·726 4·628	1·703 4·448	1·681 4·284
0·495	1·812 5·287	1·787 5·045	1·762 4·827	1·738 4·629	1·715 4·450	1·694 4·285
0·500	1·825 5·287	1·799 5·045	1·774 4·827	1·751 4·630	1·728 4·450	1·706 4·286
0·505	1·838 5·287	1·812 5·045	1·787 4·827	1·763 4·629	1·740 4·450	1·719 4·285
0·510	1·850 5·286	1·824 5·043	1·799 4·825	1·776 4·628	1·753 4·448	1·731 4·284
0·515	1·863 5·284	1·837 5·042	1·812 4·824	1·788 4·626	1·766 4·447	1·744 4·283
0·520	1·875 5·281	1·849 5·039	1·825 4·821	1·801 4·624	1·778 4·444	1·756 4·280
0·525	1·888 5·277	1·862 5·035	1·837 4·817	1·813 4·620	1·791 4·441	1·769 4·277
0·530	1·900 5·272	1·874 5·031	1·850 4·813	1·826 4·616	1·803 4·437	1·781 4·274
0·535	1·913 5·267	1·887 5·026	1·862 4·808	1·839 4·612	1·816 4·433	1·794 4·269
0·540	1·925 5·261	1·900 5·020	1·875 4·803	1·851 4·606	1·828 4·427	1·806 4·264
0·545	1·938 5·254	1·912 5·013	1·887 4·796	1·864 4·600	1·841 4·422	1·819 4·258
0·550	1·951 5·246	1·925 5·005	1·900 4·789	1·876 4·593	1·854 4·415	1·832 4·252
Cutoff	**1·825**	**1·799**	**1·774**	**1·751**	**1·728**	**1·706**

Hit Rate	0·0340	0·0360	0·0380	0·0400	0·0420	0·0440
0·555	1·963 5·237	1·937 4·997	1·913 4·781	1·889 4·585	1·866 4·408	1·844 4·245
0·560	1·976 5·227	1·950 4·988	1·925 4·772	1·902 4·577	1·879 4·399	1·857 4·237
0·565	1·989 5·217	1·963 4·978	1·938 4·763	1·914 4·568	1·892 4·391	1·870 4·229
0·570	2·001 5·206	1·975 4·967	1·951 4·752	1·927 4·558	1·904 4·381	1·882 4·220
0·575	2·014 5·194	1·988 4·956	1·964 4·741	1·940 4·547	1·917 4·371	1·895 4·210
0·580	2·027 5·181	2·001 4·943	1·976 4·730	1·953 4·536	1·930 4·360	1·908 4·199
0·585	2·040 5·167	2·014 4·930	1·989 4·717	1·965 4·524	1·943 4·348	1·921 4·188
0·590	2·053 5·152	2·027 4·916	2·002 4·704	1·978 4·511	1·955 4·336	1·934 4·176
0·595	2·065 5·137	2·040 4·901	2·015 4·689	1·991 4·498	1·968 4·323	1·946 4·164
0·600	2·078 5·120	2·052 4·886	2·028 4·675	2·004 4·483	1·981 4·309	1·959 4·150
0·605	2·091 5·103	2·065 4·869	2·041 4·659	2·017 4·468	1·994 4·295	1·972 4·136
0·610	2·104 5·085	2·078 4·852	2·054 4·642	2·030 4·452	2·007 4·280	1·985 4·122
0·615	2·117 5·066	2·091 4·834	2·067 4·625	2·043 4·436	2·020 4·264	1·998 4·106
0·620	2·130 5·046	2·105 4·815	2·080 4·607	2·056 4·418	2·033 4·247	2·012 4·090
0·625	2 144 5·026	2·118 4·795	2·093 4·588	2·069 4·400	2·047 4·230	2·025 4·074
0·630	2·157 5·004	2·131 4·775	2·106 4·568	2·083 4·381	2·060 4·211	2·038 4·056
0·635	2·170 4·982	2·144 4·753	2·120 4·548	2·096 4·362	2·073 4·193	2·051 4·038
0·640	2·183 4·958	2·158 4·731	2·133 4·527	2·109 4·341	2·086 4·173	2·065 4·019
0·645	2·197 4·934	2·171 4·708	2·146 4·505	2·123 4·320	2·100 4·153	2·078 3·999
0·650	2·210 4·909	2·184 4·684	2·160 4·482	2·136 4·298	2·113 4·132	2·091 3·979
Cutoff	1·825	1·799	1·774	1·751	1·728	1·706

Hit Rate	0·0340	0·0360	0·0380	0·0400	0·0420	0·0440
0·655	2·224	2·198	2·173	2·150	2·127	2·105
	4·883	4·659	4·458	4·276	4·110	3·958
0·660	2·237	2·212	2·187	2·163	2·140	2·119
	4·856	4·634	4·433	4·252	4·087	3·936
0·665	2·251	2·225	2·201	2·177	2·154	2·132
	4·828	4·607	4·408	4·228	4·064	3·914
0·670	2·265	2·239	2·214	2·191	2·168	2·146
	4·800	4·580	4·382	4·203	4·039	3·890
0·675	2·279	2·253	2·228	2·204	2·182	2·160
	4·770	4·552	4·355	4·177	4·015	3·866
0·680	2·293	2·267	2·242	2·218	2·196	2·174
	4·740	4·522	4·327	4·150	3·989	3·842
0·685	2·307	2·281	2·256	2·232	2·210	2·188
	4·708	4·492	4·298	4·122	3·962	3·816
0·690	2·321	2·295	2·270	2·247	2·224	2·202
	4·676	4·461	4·269	4·094	3·935	3·790
0·695	2·335	2·309	2·284	2·261	2·238	2·216
	4·642	4·430	4·238	4·065	3·907	3·763
0·700	2·349	2·324	2·299	2·275	2·252	2·230
	4·608	4·397	4·207	4·035	3·878	3·735
0·710	2·378	2·353	2·328	2·304	2·281	2·259
	4·537	4·329	4·142	3·972	3·818	3·677
0·720	2·408	2·382	2·357	2·334	2·311	2·289
	4·461	4·257	4·073	3·906	3·755	3·616
0·730	2·438	2·412	2·387	2·363	2·341	2·319
	4·382	4·181	4·001	3·837	3·688	3·552
0·740	2·468	2·442	2·418	2·394	2·371	2·349
	4·299	4·102	3·925	3·764	3·618	3·485
0·750	2·499	2·474	2·449	2·425	2·402	2·381
	4·212	4·019	3·845	3·688	3·545	3·414
0·760	2·531	2·505	2·481	2·457	2·434	2·412
	4·120	3·931	3·761	3·608	3·468	3·340
0·770	2·564	2·538	2·513	2·490	2·467	2·445
	4·024	3·840	3·674	3·524	3·387	3·262
0·780	2·597	2·571	2·547	2·523	2·500	2·478
	3·924	3·744	3·583	3·436	3·303	3·181
0·790	2·631	2·606	2·581	2·557	2·534	2·512
	3·820	3·645	3·487	3·344	3·215	3·096
0·800	2·667	2·641	2·616	2·592	2·570	2·548
	3·710	3·540	3·387	3·249	3·123	3·008
Cutoff	1·825	1·799	1·774	1·751	1·728	1·706

Hit Rate	0·0340	0·0360	0·0380	0·0400	0·0420	0·0440
0·810	2·703	2·677	2·652	2·629	2·606	2·584
	3·597	3·432	3·283	3·149	3·027	2·915
0·820	2·740	2·714	2·690	2·666	2·643	2·621
	3·478	3·318	3·175	3·045	2·927	2·819
0·830	2·779	2·753	2·729	2·705	2·682	2·660
	3·354	3·200	3·062	2·937	2·823	2·718
0·840	2·819	2·794	2·769	2·745	2·722	2·701
	3·225	3·077	2·944	2·824	2·714	2·614
0·850	2·861	2·836	2·811	2·787	2·764	2·742
	3·090	2·949	2·821	2·706	2·601	2·505
0·860	2·905	2·879	2·855	2·831	2·808	2·786
	2·950	2·815	2·693	2·583	2·483	2·391
0·870	2·951	2·926	2·901	2·877	2·854	2·832
	2·804	2·675	2·560	2·455	2·360	2·273
0·880	3·000	2·974	2·949	2·926	2·903	2·881
	2·651	2·530	2·420	2·321	2·231	2·149
0·890	3·052	3·026	3·001	2·977	2·954	2·933
	2·492	2·378	2·275	2·182	2·097	2·020
0·900	3·107	3·081	3·056	3·032	3·009	2·988
	2·326	2·219	2·123	2·037	1·958	1·885
0·910	3·166	3·140	3·115	3·091	3·069	3·047
	2·152	2·054	1·965	1·884	1·811	1·745
0·920	3·230	3·204	3·179	3·156	3·133	3·111
	1·970	1·880	1·799	1·725	1·658	1·597
0·930	3·301	3·275	3·250	3·226	3·204	3·182
	1·780	1·698	1·625	1·558	1·498	1·442
0·940	3·380	3·354	3·329	3·305	3·283	3·261
	1·579	1·506	1·441	1·382	1·329	1·280
0·950	3·470	3·444	3·419	3·396	3·373	3·351
	1·367	1·304	1·248	1·197	1·150	1 108
0·960	3·576	3·550	3·525	3·501	3·479	3·457
	1·142	1·090	1·043	1·000	0·961	0·926
0·970	3·706	3·680	3·655	3·631	3·609	3·587
	0·902	0·860	0·823	0·790	0·759	0·731
0·980	3·879	3·853	3·828	3·804	3·782	3·760
	0·642	0·612	0·586	0·562	0·540	0·520
0·990	4·151	4·125	4·101	4·077	4·054	4·032
	0·353	0·337	0·322	0·309	0·297	0·286
0·999	4·915	4·889	4·865	4·841	4·818	4·796
	0·045	0·043	0·041	0·039	0·038	0·036
Cutoff	1·825	1·799	1·774	1·751	1·728	1·706

Hit Rate	0·0460	0·0480	0·0500	0·0520	0·0540	0·0560
0·010	−0·641	−0·662	−0·681	−0·701	−0·719	−0·737
	0·276	0·267	0·258	0·250	0·243	0·236
0·020	−0·369	−0·389	−0·409	−0·428	−0·447	−0·464
	0·502	0·485	0·469	0·455	0·442	0·429
0·030	−0·196	−0·216	−0·236	−0·255	−0·274	−0·292
	0·705	0·682	0·660	0·639	0·621	0·603
0·040	−0·066	−0·086	−0·106	−0·125	−0·143	−0·161
	0·893	0·863	0·836	0·810	0·786	0·764
0·050	0·040	0·020	0·000	−0·019	−0·038	−0·056
	1·069	1·033	1·000	0·969	0·941	0·914
0·060	0·130	0·110	0·090	0·071	0·052	0·034
	1·235	1·193	1·155	1·120	1·087	1·056
0·070	0·209	0·189	0·169	0·150	0·131	0·113
	1·392	1·345	1·302	1·262	1·225	1·190
0·080	0·280	0·259	0·240	0·221	0·202	0·184
	1·541	1·489	1·441	1·397	1·356	1·318
0·090	0·344	0·324	0·304	0·285	0·266	0·249
	1·683	1·627	1·575	1·526	1·481	1·439
0·100	0·403	0·383	0·363	0·344	0·326	0·308
	1·819	1·758	1·702	1·649	1·601	1·555
0·110	0·458	0·438	0·418	0·399	0·381	0·363
	1·949	1·884	1·823	1·767	1·715	1·666
0·120	0·510	0·490	0·470	0·451	0·432	0·414
	2·073	2·004	1·940	1·880	1·825	1·773
0·130	0·559	0·538	0·518	0·499	0·481	0·463
	2·193	2·119	2·051	1·988	1·929	1·875
0·140	0·605	0·584	0·565	0·545	0·527	0·509
	2·307	2·230	2·158	2·092	2·030	1·973
0·150	0·649	0·628	0·608	0·589	0·571	0·553
	2·417	2·336	2·261	2·191	2·127	2·066
0·160	0·690	0·670	0·650	0·631	0·613	0·595
	2·522	2·437	2·359	2·287	2·219	2·156
0·170	0·731	0·710	0·691	0·672	0·653	0·635
	2·623	2·535	2·454	2·378	2·308	2·243
0·180	0·770	0·749	0·729	0·710	0·692	0·674
	2·720	2·629	2·544	2·466	2·393	2·326
0·190	0·807	0·787	0·767	0·748	0·729	0·711
	2·813	2·718	2·631	2·550	2·475	2·405
0·200	0·843	0·823	0·803	0·784	0·766	0·748
	2·902	2·804	2·715	2·631	2·553	2·481
Cutoff	1·685	1·665	1·645	1·626	1·607	1·589

Hit Rate	0·0460	0·0480	0·0500	0·0520	0·0540	0·0560
0·210	0·879	0·858	0·838	0·819	0·801	0·783
	2·987	2·887	2·794	2·709	2·629	2·554
0·220	0·913	0·892	0·873	0·854	0·835	0·817
	3·069	2·966	2·871	2·783	2·701	2·624
0·230	0·946	0·926	0·906	0·887	0·868	0·850
	3·147	3·042	2·944	2·854	2·770	2·691
0·240	0·979	0·958	0·939	0·919	0·901	0·883
	3·222	3·114	3·014	2·922	2·835	2·755
0·250	1·010	0·990	0·970	0·951	0·933	0·915
	3·294	3·183	3·081	2·986	2·898	2·816
0·260	1·042	1·021	1·002	0·982	0·964	0·946
	3·362	3·249	3·145	3·048	2·958	2·875
0·270	1·072	1·052	1·032	1·013	0·994	0·976
	3·427	3·312	3·206	3·107	3·016	2·930
0·280	1·102	1·082	1·062	1·043	1·024	1·006
	3·489	3·372	3·264	3·164	3·070	2·983
0·290	1·132	1·111	1·091	1·072	1·054	1·036
	3·548	3·429	3·319	3·217	3·122	3·034
0·300	1·161	1·140	1·120	1·101	1·083	1·065
	3·604	3·483	3·371	3·268	3·171	3·081
0·305	1·175	1·154	1·135	1·116	1·097	1·079
	3·631	3·509	3·396	3·292	3·195	3·104
0·310	1·189	1·169	1·149	1·130	1·111	1·093
	3·657	3·534	3·421	3·316	3·218	3·127
0·315	1·203	1·183	1·163	1·144	1·126	1·108
	3·682	3·559	3·444	3·339	3·240	3·148
0·320	1·217	1·197	1·177	1·158	1·140	1·122
	3·707	3·582	3·467	3·361	3·262	3·169
0·325	1·231	1·211	1·191	1·172	1·153	1·136
	3·731	3·605	3·490	3·382	3·283	3 190
0·330	1·245	1·225	1·205	1·186	1·167	1·149
	3·754	3·628	3·511	3·403	3·303	3·210
0·335	1·259	1·238	1·219	1·200	1·181	1·163
	3·776	3·649	3·532	3·424	3·323	3·229
0·340	1·272	1·252	1·232	1·213	1·195	1·177
	3·798	3·670	3·553	3·444	3·342	3·247
0·345	1·286	1·266	1·246	1·227	1·208	1·190
	3·819	3·691	3·572	3·463	3·360	3·265
0·350	1·300	1·279	1·260	1·240	1·222	1·204
	3·839	3·710	3·591	3·481	3·378	3·283
Cutoff	1·685	1·665	1·645	1·626	1·607	1·589

Hit Rate	0·0460	0·0480	0·0500	0·0520	0·0540	0·0560
0·355	1·313	1·293	1·273	1·254	1·235	1·217
	3·859	3·729	3·610	3·499	3·396	3·299
0·360	1·326	1·306	1·286	1·267	1·249	1·231
	3·878	3·748	3·627	3·516	3·412	3·316
0·365	1·340	1·319	1·300	1·281	1·262	1·244
	3·896	3·765	3·644	3·532	3·428	3·331
0·370	1·353	1·333	1·313	1·294	1·275	1·257
	3·914	3·782	3·661	3·548	3·444	3·346
0·375	1·366	1·346	1·326	1·307	1·289	1·271
	3·930	3·799	3·677	3·564	3·459	3·361
0·380	1·379	1·359	1·339	1·320	1·302	1·284
	3·947	3·814	3·692	3·578	3·473	3·374
0·385	1·393	1·372	1·352	1·333	1·315	1·297
	3·962	3·829	3·706	3·592	3·486	3·388
0·390	1·406	1·385	1·366	1·346	1·328	1·310
	3·977	3·843	3·720	3·606	3·499	3·400
0·395	1·419	1·398	1·379	1·359	1·341	1·323
	3·991	3·857	3·733	3·619	3·512	3·412
0·400	1·432	1·411	1·392	1·372	1·354	1·336
	4·005	3·870	3·746	3·631	3·524	3·424
0·405	1·445	1·424	1·404	1·385	1·367	1·349
	4·017	3·883	3·758	3·642	3·535	3·435
0·410	1·457	1·437	1·417	1·398	1·380	1·362
	4·029	3·894	3·769	3·653	3·546	3·445
0·415	1·470	1·450	1·430	1·411	1·393	1·375
	4·041	3·905	3·780	3·664	3·556	3·455
0·420	1·483	1·463	1·443	1·424	1·405	1·387
	4·052	3·916	3·790	3·674	3·565	3·464
0·425	1·496	1·475	1·456	1·437	1·418	1·400
	4·062	3·926	3·800	3·683	3·574	3·473
0·430	1·509	1·488	1·468	1·449	1·431	1·413
	4·071	3·935	3·808	3·691	3·583	3·481
0·435	1·521	1·501	1·481	1·462	1·444	1·426
	4·080	3·943	3·817	3·699	3·590	3·489
0·440	1·534	1·514	1·494	1·475	1·456	1·438
	4·088	3·951	3·824	3·707	3·597	3·496
0·445	1·547	1·526	1·507	1·487	1·469	1·451
	4·096	3·958	3·831	3·714	3·604	3·502
0·450	1·559	1·539	1·519	1·500	1·482	1·464
	4·103	3·965	3·838	3·720	3·610	3·508
Cutoff	1·685	1·665	1·645	1·626	1·607	1·589

Hit Rate	0·0460	0·0480	0·0500	0·0520	0·0540	0·0560
0·455	1·572 4·109	1·552 3·971	1·532 3·843	1·513 3·725	1·494 3·616	1·476 3·513
0·460	1·585 4·114	1·564 3·976	1·544 3·849	1·525 3·730	1·507 3·620	1·489 3·518
0·465	1·597 4·119	1·577 3·981	1·557 3·853	1·538 3·735	1·519 3·625	1·501 3·522
0·470	1·610 4·123	1·589 3·985	1·570 3·857	1·550 3·739	1·532 3·628	1·514 3·526
0·475	1·622 4·127	1·602 3·989	1·582 3·861	1·563 3·742	1·545 3·632	1·527 3·529
0·480	1·635 4·130	1·614 3·991	1·595 3·863	1·576 3·745	1·557 3·634	1·539 3·531
0·485	1·647 4·132	1·627 3·994	1·607 3·865	1·588 3·747	1·570 3·636	1·552 3·533
0·490	1·660 4·134	1·639 3·995	1·620 3·867	1·601 3·748	1·582 3·638	1·564 3·534
0·495	1·672 4·135	1·652 3·996	1·632 3·868	1·613 3·749	1·595 3·638	1·577 3·535
0·500	1·685 4·135	1·665 3·996	1·645 3·868	1·626 3·749	1·607 3·639	1·589 3·536
0·505	1·697 4·135	1·677 3·996	1·657 3·868	1·638 3·749	1·620 3·638	1·602 3·535
0·510	1·710 4·134	1·690 3·995	1·670 3·867	1·651 3·748	1·632 3·638	1·614 3·534
0·515	1·723 4·132	1·702 3·994	1·682 3·865	1·663 3·747	1·645 3·636	1·627 3·533
0·520	1·735 4·130	1·715 3·991	1·695 3·863	1·676 3·745	1·657 3·634	1·639 3·531
0·525	1·748 4·127	1·727 3·989	1·708 3·861	1·688 3·742	1·670 3·632	1·652 3·529
0·530	1·760 4·123	1·740 3·985	1·720 3·857	1·701 3·739	1·683 3·628	1·665 3·526
0·535	1·773 4·119	1·752 3·981	1·733 3·853	1·714 3·735	1·695 3·625	1·677 3·522
0·540	1·785 4·114	1·765 3·976	1·745 3·849	1·726 3·730	1·708 3·620	1·690 3·518
0·545	1·798 4·109	1·778 3·971	1·758 3·843	1·739 3·725	1·720 3·616	1·702 3·513
0·550	1·811 4·103	1·790 3·965	1·771 3·838	1·751 3·720	1·733 3·610	1·715 3·508
Cutoff	1·685	1·665	1·645	1·626	1·607	1·589

Hit Rate	0·0460	0·0480	0·0500	0·0520	0·0540	0·0560
0·555	1·823 4·096	1·803 3·958	1·783 3·831	1·764 3·714	1·746 3·604	1·728 3·502
0·560	1·836 4·088	1·816 3·951	1·796 3·824	1·777 3·707	1·758 3·597	1·740 3·496
0·565	1·849 4·080	1·828 3·943	1·809 3·817	1·789 3·699	1·771 3·590	1·753 3·489
0·570	1·861 4·071	1·841 3·935	1·821 3·808	1·802 3·691	1·784 3·583	1·766 3·481
0·575	1·874 4·062	1·854 3·926	1·834 3·800	1·815 3·683	1·796 3·574	1·778 3·473
0·580	1·887 4·052	1·866 3·916	1·847 3·790	1·828 3·674	1·809 3·565	1·791 3·464
0·585	1·900 4·041	1·879 3·905	1·860 3·780	1·840 3·664	1·822 3·556	1·804 3·455
0·590	1·912 4·029	1·892 3·894	1·872 3·769	1·853 3·653	1·835 3·546	1·817 3·445
0·595	1·925 4·017	1·905 3·883	1·885 3·758	1·866 3·642	1·848 3·535	1·830 3·435
0·600	1·938 4·005	1·918 3·870	1·898 3·746	1·879 3·631	1·861 3·524	1·843 3·424
0·605	1·951 3·991	1·931 3·857	1·911 3·733	1·892 3·619	1·874 3·512	1·856 3·412
0·610	1·964 3·977	1·944 3·843	1·924 3·720	1·905 3·606	1·887 3·499	1·869 3·400
0·615	1·977 3·962	1·957 3·829	1·937 3·706	1·918 3·592	1·900 3·486	1·882 3·388
0·620	1·990 3·947	1·970 3·814	1·950 3·692	1·931 3·578	1·913 3·473	1·895 3·374
0·625	2·004 3·930	1·983 3·799	1·963 3·677	1·944 3·564	1·926 3·459	1·908 3·361
0·630	2·017· 3·914	1·996 3·782	1·977 3·661	1·958 3·548	1·939 3·444	1·921 3·346
0·635	2·030 3·896	2·010 3·765	1·990 3·644	1·971 3·532	1·952 3·428	1·934 3·331
0·640	2·043 3·878	2·023 3·748	2·003 3·627	1·984 3·516	1·966 3·412	1·948 3·316
0·645	2·057 3·859	2·036 3·729	2·017 3·610	1·998 3·499	1·979 3·396	1·961 3·299
0·650	2·070 3·839	2·050 3·710	2·030 3·591	2·011 3·481	1·993 3·378	1·975 3·283
Cutoff	1·685	1·665	1·645	1·626	1·607	1·589

Hit Rate	0·0460	0·0480	0·0500	0·0520	0·0540	0·0560
0·655	2·084 3·819	2·063 3·691	2·044 3·572	2·025 3·463	2·006 3·360	1·988 3·265
0·660	2·097 3·798	2·077 3·670	2·057 3·553	2·038 3·444	2·020 3·342	2·002 3·247
0·665	2·111 3·776	2·091 3·649	2·071 3·532	2·052 3·424	2·033 3·323	2·015 3·229
0·670	2·125 3·754	2·104 3·628	2·085 3·511	2·066 3·403	2·047 3·303	2·029 3·210
0·675	2·139 3·731	2·118 3·605	2·099 3·490	2·080 3·382	2·061 3·283	2·043 3·190
0·680	2·153 3·707	2·132 3·582	2·113 3·467	2·093 3·361	2·075 3·262	2·057 3·169
0·685	2·167 3·682	2·146 3·559	2·127 3·444	2·107 3·339	2·089 3·240	2·071 3·148
0·690	2·181 3·657	2·160 3·534	2·141 3·421	2·122 3·316	2·103 3·218	2·085 3·127
0·695	2·195 3·631	2·175 3·509	2·155 3·396	2·136 3·292	2·117 3·195	2·099 3·104
0·700	2·209 3·604	2·189 3·483	2·169 3·371	2·150 3·268	2·132 3·171	2·114 3·081
0·710	2·238 3·548	2·218 3·429	2·198 3·319	2·179 3·217	2·161 3·122	2·143 3·034
0·720	2·268 3·489	2·247 3·372	2·228 3·264	2·209 3·164	2·190 3·070	2·172 2·983
0·730	2·298 3·427	2·277 3·312	2·258 3·206	2·239 3·107	2·220 3·016	2·202 2·930
0·740	2·328 3·362	2·308 3·249	2·288 3·145	2·269 3·048	2·251 2·958	2·233 2·875
0·750	2·359 3·294	2·339 3·183	2·319 3·081	2·300 2·986	2·282 2·898	2·264 2·816
0·760	2·391 3·222	2·371 3·114	2·351 3·014	2·332 2·922	2·314 2·835	2·296 2·755
0·770	2·424 3·147	2·403 3·042	2·384 2·944	2·365 2·854	2·346 2·770	2·328 2·691
0·780	2·457 3·069	2·437 2·966	2·417 2·871	2·398 2·783	2·379 2·701	2·361 2·624
0·790	2·491 2·987	2·471 2·887	2·451 2·794	2·432 2·709	2·414 2·629	2·396 2·554
0·800	2·527 2·902	2·506 2·804	2·486 2·715	2·467 2·631	2·449 2·553	2·431 2·481
Cutoff	1·685	1·665	1·645	1·626	1·607	1·589

FALSE POSITIVE RATE

Hit Rate	0·0460	0·0480	0·0500	0·0520	0·0540	0·0560
0·810	2·563 2·813	2·542 2·718	2·523 2·631	2·504 2·550	2·485 2·475	2·467 2·405
0·820	2·600 2·720	2·580 2·629	2·560 2·544	2·541 2·466	2·523 2·393	2·505 2·326
0·830	2·639 2·623	2·619 2·535	2·599 2·454	2·580 2·378	2·561 2·308	2·543 2·243
0·840	2·679 2·522	2·659 2·437	2·639 2·359	2·620 2·287	2·602 2·219	2·584 2·156
0·850	2·721 2·417	2·701 2·336	2·681 2·261	2·662 2·191	2·644 2·127	2·626 2·066
0·860	2·765 2·307	2·745 2·230	2·725 2·158	2·706 2·092	2·688 2·030	2·670 1·973
0·870	2·811 2·193	2·791 2·119	2·771 2·051	2·752 1·988	2·734 1·929	2·716 1·875
0·880	2·860 2·073	2·840 2·004	2·820 1·940	2·801 1·880	2·782 1·825	2·764 1·773
0·890	2·911 1·949	2·891 1·884	2·871 1·823	2·852 1·767	2·834 1·715	2·816 1·666
0·900	2·966 1·819	2·946 1·758	2·926 1·702	2·907 1·649	2·889 1·601	2·871 1·555
0·910	3·026 1·683	3·005 1·627	2·986 1·575	2·967 1·526	2·948 1·481	2·930 1·439
0·920	3·090 1·541	3·070 1·489	3·050 1·441	3·031 1·397	3·012 1·356	2·994 1·318
0·930	3·161 1·392	3·140 1·345	3·121 1·302	3·102 1·262	3·083 1·225	3·065 1·190
0·940	3·240 1·235	3·219 1·193	3·200 1·155	3·181 1·120	3·162 1·087	3·144 1·056
0·950	3·330 1·069	3·309 1·033	3·290 1·000	3·271 0·969	3·252 0·941	3·234 0·914
0·960	3·436 0·893	3·415 0·863	3·396 0·836	3·376 0·810	3·358 0·786	3·340 0·764
0·970	3·566 0·705	3·545 0·682	3·526 0·660	3·507 0·639	3·488 0·621	3·470 0·603
0·980	3·739 0·502	3·718 0·485	3·699 0·469	3·680 0·455	3·661 0·442	3·643 0·429
0·990	4·011 0·276	3·991 0·267	3·971 0·258	3·952 0·250	3·934 0·243	3·916 0·236
0·999	4·775 0·035	4·755 0·034	4·735 0·033	4·716 0·032	4·697 0·031	4·679 0·030
Cutoff	1·685	1·665	1·645	1·626	1·607	1·589

Hit Rate	0·0580	0·0600	0·0620	0·0640	0·0660	0·0680
0·010	−0·755	−0·772	−0·788	−0·804	−0·820	−0·835
	0·230	0·224	0·218	0·213	0·208	0·203
0·020	−0·482	−0·499	−0·516	−0·532	−0·547	−0·563
	0·417	0·406	0·396	0·386	0·377	0·369
0·030	−0·309	−0·326	−0·343	−0·359	−0·375	−0·390
	0·587	0·571	0·557	0·543	0·530	0·518
0·040	−0·179	−0·196	−0·212	−0·229	−0·244	−0·260
	0·743	0·723	0·705	0·688	0·672	0·656
0·050	−0·073	−0·090	−0·107	−0·123	−0·139	−0·154
	0·889	0·866	0·844	0·823	0·804	0·785
0·060	0·017	0·000	−0·017	−0·033	−0·049	−0·064
	1·027	1·000	0·975	0·951	0·928	0·907
0·070	0·096	0·079	0·062	0·046	0·030	0·015
	1·158	1·127	1·099	1·072	1·046	1·023
0·080	0·167	0·150	0·133	0·117	0·101	0·086
	1·282	1·248	1·216	1·187	1·159	1·132
0·090	0·231	0·214	0·197	0·181	0·166	0·150
	1·400	1·363	1·329	1·296	1·266	1·237
0·100	0·290	0·273	0·257	0·240	0·225	0·209
	1·513	1·473	1·436	1·401	1·368	1·337
0·110	0·345	0·328	0·312	0·296	0·280	0·264
	1·621	1·578	1·539	1·501	1·466	1·432
0·120	0·397	0·380	0·363	0·347	0·331	0·316
	1·725	1·679	1·637	1·597	1·559	1·524
0·130	0·445	0·428	0·412	0·396	0·380	0·364
	1·824	1·776	1·731	1·689	1·649	1·611
0·140	0·491	0·474	0·458	0·442	0·426	0·411
	1·919	1·868	1·821	1·777	1·735	1·695
0·150	0·535	0·518	0·502	0·486	0·470	0·454
	2·010	1·957	1·908	1·861	1·817	1·776
0·160	0·577	0·560	0·544	0·528	0·512	0·496
	2·098	2·043	1·991	1·942	1·896	1·853
0·170	0·618	0·601	0·584	0·568	0·552	0·537
	2·182	2·124	2·071	2·020	1·972	1·927
0·180	0·656	0·639	0·623	0·607	0·591	0·575
	2·262	2·203	2·147	2·095	2·045	1·998
0·190	0·694	0·677	0·660	0·644	0·628	0·613
	2·339	2·278	2·220	2·166	2·115	2·067
0·200	0·730	0·713	0·697	0·680	0·665	0·649
	2·414	2·350	2·291	2·235	2·182	2·132
Cutoff	1·572	1·555	1·538	1·522	1·506	1·491

FALSE POSITIVE RATE

Hit Rate	0·0580	0·0600	0·0620	0·0640	0·0660	0·0680
0·210	0·765 2·485	0·748 2·419	0·732 2·358	0·716 2·301	0·700 2·246	0·684 2·195
0·220	0·800 2·553	0·783 2·486	0·766 2·423	0·750 2·364	0·734 2·308	0·719 2·255
0·230	0·833 2·618	0·816 2·549	0·799 2·485	0·783 2·424	0·767 2·367	0·752 2·313
0·240	0·865 2·680	0·848 2·610	0·832 2·544	0·816 2·482	0·800 2·423	0·785 2·368
0·250	0·897 2·740	0·880 2·668	0·864 2·600	0·848 2·537	0·832 2·477	0·816 2·420
0·260	0·928 2·796	0·911 2·723	0·895 2·654	0·879 2·589	0·863 2·528	0·848 2·470
0·270	0·959 2·850	0·942 2·776	0·925 2·705	0·909 2·639	0·893 2·577	0·878 2·518
0·280	0·989 2·902	0·972 2·826	0·955 2·754	0·939 2·687	0·923 2·624	0·908 2·564
0·290	1·018 2·951	1·001 2·874	0·985 2·801	0·969 2·732	0·953 2·668	0·937 2·607
0·300	1·047 2·997	1·030 2·919	1·014 2·845	0·998 2·775	0·982 2·710	0·966 2·648
0·305	1·062 3·020	1·045 2·940	1·028 2·866	1·012 2·796	0·996 2·730	0·981 2·668
0·310	1·076 3·041	1·059 2·962	1·042 2·887	1·026 2·816	1·010 2·750	0·995 2·687
0·315	1·090 3·062	1·073 2·982	1·056 2·907	1·040 2·836	1·025 2·769	1·009 2·706
0·320	1·104 3·083	1·087 3·002	1·071 2·926	1·054 2·855	1·039 2·787	1·023 2·724
0·325	1·118 3·103	1·101 3·021	1·084 2·945	1·068 2·873	1·052 2·805	1·037 2·741
0·330	1·132 3·122	1·115 3·040	1·098 2·963	1·082 2·891	1·066 2·823	1·051 2·758
0·335	1·146 3·141	1·129 3·058	1·112 2·981	1·096 2·908	1·080 2·839	1·065 2·775
0·340	1·159 3·159	1·142 3·076	1·126 2·998	1·110 2·925	1·094 2·856	1·078 2·791
0·345	1·173 3·176	1·156 3·093	1·139 3·015	1·123 2·941	1·107 2·872	1·092 2·806
0·350	1·186 3·193	1·169 3·109	1·153 3·031	1·137 2·957	1·121 2·887	1·106 2·821
Cutoff	1·572	1·555	1·538	1·522	1·506	1·491

Hit Rate	0·0580	0·0600	0·0620	0·0640	0·0660	0·0680
0·355	1·200 3·210	1·183 3·125	1·166 3·046	1·150 2·972	1·134 2·902	1·119 2·835
0·360	1·213 3·225	1·196 3·141	1·180 3·061	1·164 2·986	1·148 2·916	1·132 2·849
0·365	1·227 3·240	1·210 3·155	1·193 3·076	1·177 3·000	1·161 2·930	1·146 2·863
0·370	1·240 3·255	1·223 3·170	1·206 3·089	1·190 3·014	1·174 2·943	1·159 2·876
0·375	1·253 3·269	1·236 3·183	1·220 3·103	1·203 3·027	1·188 2·955	1·172 2·888
0·380	1·266 3·282	1·249 3·196	1·233 3·115	1·217 3·039	1·201 2·968	1·185 2·900
0·385	1·279 3·295	1·262 3·209	1·246 3·128	1·230 3·051	1·214 2·979	1·198 2·911
0·390	1·292 3·308	1·275 3·221	1·259 3·139	1·243 3·063	1·227 2·990	1·212 2·922
0·395	1·305 3·319	1·288 3·232	1·272 3·151	1·256 3·074	1·240 3·001	1·225 2·933
0·400	1·318 3·331	1·301 3·243	1·285 3·161	1·269 3·084	1·253 3·011	1·238 2·942
0·405	1·331 3·341	1·314 3·254	1·298 3·171	1·282 3·094	1·266 3·021	1·250 2·952
0·410	1·344 3·351	1·327 3·263	1·311 3·181	1·294 3·103	1·279 3·030	1·263 2·961
0·415	1·357 3·361	1·340 3·273	1·323 3·190	1·307 3·112	1·292 3·038	1·276 2·969
0·420	1·370 3·370	1·353 3·281	1·336 3·198	1·320 3·120	1·304 3·047	1·289 2·977
0·425	1·383 3·378	1·366 3·290	1·349 3·206	1·333 3·128	1·317 3·054	1·302 2 985
0·430	1·395 3·386	1·378 3·297	1·362 3·214	1·346 3·135	1·330 3·061	1·314 2·991
0·435	1·408 3·394	1·391 3·304	1·375 3·221	1·358 3·142	1·343 3·068	1·327 2·998
0·440	1·421 3·400	1·404 3·311	1·387 3·227	1·371 3·148	1·355 3·074	1·340 3·004
0·445	1·433 3·407	1·416 3·317	1·400 3·233	1·384 3·154	1·368 3·080	1·353 3·009
0·450	1·446 3·412	1·429 3·323	1·413 3·239	1·396 3·159	1·381 3·085	1·365 3·014
Cutoff	1·572	1·555	1·538	1·522	1·506	1·491

Hit Rate	0·0580	0·0600	0·0620	0·0640	0·0660	0·0680
0·455	1·459 3·417	1·442 3·328	1·425 3·243	1·409 3·164	1·393 3·090	1·378 3·019
0·460	1·471 3·422	1·454 3·332	1·438 3·248	1·422 3·168	1·406 3·094	1·390 3·023
0·465	1·484 3·426	1·467 3·336	1·450 3·252	1·434 3·172	1·418 3·097	1·403 3·027
0·470	1·497 3·430	1·480 3·340	1·463 3·255	1·447 3·175	1·431 3·101	1·416 3·030
0·475	1·509 3·433	1·492 3·342	1·475 3·258	1·459 3·178	1·444 3·103	1·428 3·032
0·480	1·522 3·435	1·505 3·345	1·488 3·260	1·472 3·181	1·456 3·105	1·441 3·035
0·485	1·534 3·437	1·517 3·347	1·501 3·262	1·484 3·182	1·469 3·107	1·453 3·036
0·490	1·547 3·438	1·530 3·348	1·513 3·263	1·497 3·184	1·481 3·108	1·466 3·037
0·495	1·559 3·439	1·542 3·349	1·526 3·264	1·510 3·184	1·494 3·109	1·478 3·038
0·500	1·572 3·439	1·555 3·349	1·538 3·264	1·522 3·185	1·506 3·109	1·491 3·038
0·505	1·584 3·439	1·567 3·349	1·551 3·264	1·535 3·184	1·519 3·109	1·503 3·038
0·510	1·597 3·438	1·580 3·348	1·563 3·263	1·547 3·184	1·531 3·108	1·516 3·037
0·515	1·609 3·437	1·592 3·347	1·576 3·262	1·560 3·182	1·544 3·107	1·528 3·036
0·520	1·622 3·435	1·605 3·345	1·588 3·260	1·572 3·181	1·556 3·105	1·541 3·035
0·525	1·634 3·433	1·617 3·342	1·601 3·258	1·585 3·178	1·569 3·103	1·554 3·032
0·530	1·647 3·430	1·630 3·340	1·613 3·255	1·597 3·175	1·582 3·101	1·566 3·030
0·535	1·660 3·426	1·643 3·336	1·626 3·252	1·610 3·172	1·594 3·097	1·579 3·027
0·540	1·672 3·422	1·655 3·332	1·639 3·248	1·622 3·168	1·607 3·094	1·591 3·023
0·545	1·685 3·417	1·668 3·328	1·651 3·243	1·635 3·164	1·619 3·090	1·604 3·019
0·550	1·697 3·412	1·680 3·323	1·664 3·239	1·648 3·159	1·632 3·085	1·617 3·014
Cutoff	1·572	1·555	1·538	1·522	1·506	1·491

Hit Rate	0·0580	0·0600	0·0620	0·0640	0·0660	0·0680
0·555	1·710 3·407	1·693 3·317	1·677 3·233	1·660 3·154	1·645 3·080	1·629 3·009
0·560	1·723 3·400	1·706 3·311	1·689 3·227	1·673 3·148	1·657 3·074	1·642 3·004
0·565	1·735 3·394	1·718 3·304	1·702 3·221	1·686 3·142	1·670 3·068	1·655 2·998
0·570	1·748 3·386	1·731 3·297	1·715 3·214	1·698 3·135	1·683 3·061	1·667 2·991
0·575	1·761 3·378	1·744 3·290	1·727 3·206	1·711 3·128	1·695 3·054	1·680 2·985
0·580	1·774 3·370	1·757 3·281	1·740 3·198	1·724 3·120	1·708 3·047	1·693 2·977
0·585	1·786 3·361	1·769 3·273	1·753 3·190	1·737 3·112	1·721 3·038	1·706 2·969
0·590	1·799 3·351	1·782 3·263	1·766 3·181	1·750 3·103	1·734 3·030	1·718 2·961
0·595	1·812 3·341	1·795 3·254	1·779 3·171	1·762 3·094	1·747 3·021	1·731 2·952
0·600	1·825 3·331	1·808 3·243	1·792 3·161	1·775 3·084	1·760 3·011	1·744 2·942
0·605	1·838 3·319	1·821 3·232	1·805 3·151	1·788 3·074	1·773 3·001	1·757 2·933
0·610	1·851 3·308	1·834 3·221	1·818 3·139	1·801 3·063	1·786 2·990	1·770 2·922
0·615	1·864 3·295	1·847 3·209	1·831 3·128	1·814 3·051	1·799 2·979	1·783 2·911
0·620	1·877 3·282	1·860 3·196	1·844 3·115	1·828 3·039	1·812 2·968	1·796 2·900
0·625	1·890 3·269	1·873 3·183	1·857 3·103	1·841 3·027	1·825 2·955	1·809 2·888
0·630	1·904 3·255	1·887 3·170	1·870 3·089	1·854 3·014	1·838 2·943	1·823 2·876
0·635	1·917 3·240	1·900 3·155	1·883 3·076	1·867 3·000	1·851 2·930	1·836 2·863
0·640	1·930 3·225	1·913 3·141	1·897 3·061	1·880 2·986	1·865 2·916	1·849 2·849
0·645	1·944 3·210	1·927 3·125	1·910 3·046	1·894 2·972	1·878 2·902	1·863 2·835
0·650	1·957 3·193	1·940 3·109	1·924 3·031	1·907 2·957	1·892 2·887	1·876 2·821
Cutoff	1·572	1·555	1·538	1·522	1·506	1·491

Hit Rate	0-0580	0-0600	0-0620	0-0640	0-0660	0-0680
0-655	1·971	1·954	1·937	1·921	1·905	1·890
	3·176	3·093	3·015	2·941	2·872	2·806
0-660	1·984	1·967	1·951	1·934	1·919	1·903
	3·159	3·076	2·998	2·925	2·856	2·791
0-665	1·998	1·981	1·964	1·948	1·932	1·917
	3·141	3·058	2·981	2·908	2·839	2·775
0-670	2·012	1·995	1·978	1·962	1·946	1·931
	3·122	3·040	2·963	2·891	2·823	2·758
0-675	2·026	2·009	1·992	1·976	1·960	1·945
	3·103	3·021	2·945	2·873	2·805	2·741
0-680	2·039	2·022	2·006	1·990	1·974	1·959
	3·083	3·002	2·926	2·855	2·787	2·724
0-685	2·054	2·037	2·020	2·004	1·988	1·973
	3·062	2·982	2·907	2·836	2·769	2·706
0-690	2·068	2·051	2·034	2·018	2·002	1·987
	3·041	2·962	2·887	2·816	2·750	2·687
0-695	2·082	2·065	2·048	2·032	2·016	2·001
	3·020	2·940	2·866	2·796	2·730	2·668
0-700	2·096	2·079	2·063	2·046	2·031	2·015
	2·997	2·919	2·845	2·775	2·710	2·648
0-710	2·125	2·108	2·092	2·075	2·060	2·044
	2·951	2·874	2·801	2·732	2·668	2·607
0-720	2·155	2·138	2·121	2·105	2·089	2·074
	2·902	2·826	2·754	2·687	2·624	2·564
0-730	2·185	2·168	2·151	2·135	2·119	2·104
	2·850	2·776	2·705	2·639	2·577	2·518
0-740	2·215	2·198	2·182	2·165	2·150	2·134
	2·796	2·723	2·654	2·589	2·528	2·470
0-750	2·246	2·229	2·213	2·197	2·181	2·165
	2·740	2·668	2·600	2·537	2·477	2·420
0-760	2·278	2·261	2·245	2·228	2·213	2·197
	2·680	2·610	2·544	2·482	2·423	2·368
0-770	2·311	2·294	2·277	2·261	2·245	2·230
	2·618	2·549	2·485	2·424	2·367	2·313
0-780	2·344	2·327	2·310	2·294	2·278	2·263
	2·553	2·486	2·423	2·364	2·308	2·255
0-790	2·378	2·361	2·345	2·328	2·313	2·297
	2·485	2·419	2·358	2·301	2·246	2·195
0-800	2·413	2·396	2·380	2·364	2·348	2·332
	2·414	2·350	2·291	2·235	2·182	2·132
Cutoff	1·572	1·555	1·538	1·522	1·506	1·491

Hit Rate	0·0580	0·0600	0·0620	0·0640	0·0660	0·0680
0·810	2·450 2·339	2·433 2·278	2·416 2·220	2·400 2·166	2·384 2·115	2·369 2·067
0·820	2·487 2·262	2·470 2·203	2·454 2·147	2·437 2·095	2·422 2·045	2·406 1·998
0·830	2·526 2·182	2·509 2·124	2·492 2·071	2·476 2·020	2·460 1·972	2·445 1·927
0·840	2·566 2·098	2·549 2·043	2·533 1·991	2·516 1·942	2·501 1·896	2·485 1·853
0·850	2·608 2·010	2·591 1·957	2·575 1·908	2·558 1·861	2·543 1·817	2·527 1·776
0·860	2·652 1·919	2·635 1·868	2·619 1·821	2·602 1·777	2·587 1·735	2·571 1·695
0·870	2·698 1·824	2·681 1·776	2·665 1·731	2·648 1·689	2·633 1·649	2·617 1·611
0·880	2·747 1·725	2·730 1·679	2·713 1·637	2·697 1·597	2·681 1·559	2·666 1·524
0·890	2·798 1·621	2·781 1·578	2·765 1·539	2·749 1·501	2·733 1·466	2·717 1·432
0·900	2·853 1·513	2·836 1·473	2·820 1·436	2·804 1·401	2·788 1·368	2·772 1·337
0·910	2·913 1·400	2·896 1·363	2·879 1·329	2·863 1·296	2·847 1·266	2·832 1·237
0·920	2·977 1·282	2·960 1·248	2·943 1·216	2·927 1·187	2·911 1·159	2·896 1·132
0·930	3·048 1·158	3·031 1·127	3·014 1·099	2·998 1·072	2·982 1·046	2·967 1·023
0·940	3·127 1·027	3·110 1·000	3·093 0·975	3·077 0·951	3·061 0·928	3·046 0·907
0·950	3·217 0·889	3·200 0·866	3·183 0·844	3·167 0·823	3·151 0·804	3·136 0·785
0·960	3·322 0·743	3·305 0·723	3·289 0·705	3·273 0·688	3·257 0·672	3·242 0·656
0·970	3·453 0·587	3·436 0·571	3·419 0·557	3·403 0·543	3·387 0·530	3·372 0·518
0·980	3·626 0·417	3·609 0·406	3·592 0·396	3·576 0·386	3·560 0·377	3·545 0·369
0·990	3·898 0·230	3·881 0·224	3·865 0·218	3·848 0·213	3·833 0·208	3·817 0·203
0·999	4·662 0·029	4·645 0·028	4·628 0·028	4·612 0·027	4·596 0·026	4·581 0·026
Cutoff	1·572	1·555	1·538	1·522	1·506	1·491

Hit Rate	0·0700	0·0720	0·0740	0·0760	0·0780	0·0800
0·010	−0·851 0·198	−0·865 0·194	−0·880 0·190	−0·894 0·186	−0·908 0·183	−0·921 0·179
0·020	−0·578 0·361	−0·593 0·353	−0·607 0·346	−0·621 0·339	−0·635 0·332	−0·649 0·326
0·030	−0·405 0·507	−0·420 0·496	−0·434 0·486	−0·448 0·476	−0·462 0·467	−0·476 0·458
0·040	−0·275 0·642	−0·290 0·628	−0·304 0·615	−0·318 0·603	−0·332 0·591	−0·346 0·580
0·050	−0·169 0·768	−0·184 0·752	−0·198 0·736	−0·212 0·721	−0·226 0·707	−0·240 0·694
0·060	−0·079 0·887	−0·094 0·868	−0·108 0·850	−0·122 0·833	−0·136 0·817	−0·150 0·801
0·070	0·000 1·000	−0·015 0·979	−0·029 0·958	−0·043 0·939	−0·057 0·921	−0·071 0·903
0·080	0·071 1·107	0·056 1·084	0·042 1·061	0·027 1·040	0·014 1·019	0·000 1·000
0·090	0·135 1·209	0·120 1·184	0·106 1·159	0·092 1·136	0·078 1·113	0·064 1·092
0·100	0·194 1·307	0·180 1·279	0·165 1·253	0·151 1·227	0·137 1·203	0·124 1·180
0·110	0·249 1·400	0·235 1·370	0·220 1·342	0·206 1·315	0·192 1·289	0·179 1·265
0·120	0·301 1·490	0·286 1·458	0·272 1·428	0·258 1·399	0·244 1·372	0·230 1·346
0·130	0·349 1·576	0·335 1·542	0·320 1·510	0·306 1·479	0·292 1·451	0·279 1·423
0·140	0·395 1·658	0·381 1·622	0·366 1·589	0·352 1·557	0·338 1·526	0·325 1·497
0·150	0·439 1·737	0·425 1·699	0·410 1·664	0·396 1·631	0·382 1·599	0·369 1·568
0·160	0·481 1·812	0·467 1·773	0·452 1·737	0·438 1·702	0·424 1·668	0·411 1·637
0·170	0·522 1·885	0·507 1·844	0·492 1·806	0·478 1·770	0·464 1·735	0·451 1·702
0·180	0·560 1·954	0·546 1·912	0·531 1·873	0·517 1·835	0·503 1·799	0·490 1·765
0·190	0·598 2·021	0·583 1·978	0·569 1·937	0·555 1·898	0·541 1·861	0·527 1·825
0·200	0·634 2·085	0·619 2·040	0·605 1·998	0·591 1·958	0·577 1·920	0·563 1·883
Cutoff	1·476	1·461	1·447	1·433	1·419	1·405

Hit Rate	0·0700	0·0720	0·0740	0·0760	0·0780	0·0800
0·210	0·669	0·655	0·640	0·626	0·612	0·599
	2·146	2·101	2·057	2·016	1·976	1·939
0·220	0·704	0·689	0·674	0·660	0·646	0·633
	2·205	2·158	2·113	2·071	2·030	1·992
0·230	0·737	0·722	0·708	0·694	0·680	0·666
	2·262	2·213	2·167	2·124	2·082	2·042
0·240	0·769	0·755	0·740	0·726	0·712	0·699
	2·315	2·266	2·219	2·174	2·132	2·091
0·250	0·801	0·787	0·772	0·758	0·744	0·731
	2·367	2·316	2·268	2·222	2·179	2·138
0·260	0·832	0·818	0·803	0·789	0·775	0·762
	2·416	2·364	2·315	2·268	2·224	2·182
0·270	0·863	0·848	0·834	0·820	0·806	0·792
	2·463	2·410	2·360	2·312	2·267	2·224
0·280	0·893	0·878	0·864	0·850	0·836	0·822
	2·507	2·453	2·403	2·354	2·308	2·264
0·290	0·922	0·908	0·893	0·879	0·865	0·852
	2·549	2·495	2·443	2·394	2·347	2·303
0·300	0·951	0·937	0·922	0·908	0·894	0·881
	2·590	2·534	2·482	2·432	2·384	2·339
0·305	0·966	0·951	0·937	0·922	0·909	0·895
	2·609	2·553	2·500	2·450	2·402	2·356
0·310	0·980	0·965	0·951	0·937	0·923	0·909
	2·628	2·571	2·518	2·467	2·419	2·373
0·315	0·994	0·979	0·965	0·951	0·937	0·923
	2·646	2·589	2·535	2·484	2·436	2·389
0·320	1·008	0·993	0·979	0·965	0·951	0·937
	2·663	2·606	2·552	2·501	2·452	2·405
0·325	1·022	1·007	0·993	0·979	0·965	0·951
	2·681	2·623	2·569	2·517	2·468	2·421
0·330	1·036	1·021	1·007	0·993	0·979	0·965
	2·697	2·639	2·585	2·533	2·483	2·436
0·335	1·050	1·035	1·020	1·006	0·993	0·979
	2·713	2·655	2·600	2·548	2·498	2·451
0·340	1·063	1·049	1·034	1·020	1·006	0·993
	2·729	2·671	2·615	2·562	2·512	2·465
0·345	1·077	1·062	1·048	1·034	1·020	1·006
	2·744	2·685	2·630	2·577	2·526	2·478
0·350	1·090	1·076	1·061	1·047	1·033	1·020
	2·759	2·700	2·644	2·590	2·540	2·491
Cutoff	**1·476**	**1·461**	**1·447**	**1·433**	**1·419**	**1·405**

Hit Rate	0·0700	0·0720	0·0740	0·0760	0·0780	0·0800
0·355	1·104 2·773	1·089 2·713	1·075 2·657	1·061 2·604	1·047 2·553	1·033 2·504
0·360	1·117 2·786	1·103 2·727	1·088 2·670	1·074 2·616	1·060 2·565	1·047 2·516
0·365	1·131 2·799	1·116 2·740	1·102 2·683	1·087 2·629	1·074 2·577	1·060 2·528
0·370	1·144 2·812	1·129 2·752	1·115 2·695	1·101 2·641	1·087 2·589	1·073 2·540
0·375	1·157 2·824	1·142 2·764	1·128 2·706	1·114 2·652	1·100 2·600	1·086 2·551
0·380	1·170 2·836	1·156 2·775	1·141 2·718	1·127 2·663	1·113 2·611	1·100 2·561
0·385	1·183 2·847	1·169 2·786	1·154 2·728	1·140 2·673	1·126 2·621	1·113 2·571
0·390	1·196 2·858	1·182 2·796	1·167 2·738	1·153 2·683	1·139 2·631	1·126 2·581
0·395	1·209 2·868	1·195 2·806	1·180 2·748	1·166 2·693	1·152 2·640	1·139 2·590
0·400	1·222 2·877	1·208 2·816	1·193 2·757	1·179 2·702	1·165 2·649	1·152 2·599
0·405	1·235 2·887	1·221 2·825	1·206 2·766	1·192 2·710	1·178 2·658	1·165 2·607
0·410	1·248 2·895	1·234 2·833	1·219 2·775	1·205 2·719	1·191 2·666	1·178 2·615
0·415	1·261 2·904	1·246 2·841	1·232 2·782	1·218 2·726	1·204 2·673	1·190 2·622
0·420	1·274 2·911	1·259 2·849	1·245 2·790	1·231 2·734	1·217 2·680	1·203 2·629
0·425	1·287 2·919	1·272 2·856	1·258 2·797	1·243 2·741	1·230 2·687	1·216 2·636
0·430	1·299 2·925	1·285 2·863	1·270 2·803	1·256 2·747	1·242 2·693	1·229 2·642
0·435	1·312 2·932	1·297 2·869	1·283 2·809	1·269 2·753	1·255 2·699	1·241 2·648
0·440	1·325 2·938	1·310 2·875	1·296 2·815	1·282 2·758	1·268 2·704	1·254 2·653
0·445	1·337 2·943	1·323 2·880	1·308 2·820	1·294 2·763	1·280 2·709	1·267 2·658
0·450	1·350 2·948	1·335 2·885	1·321 2·825	1·307 2·768	1·293 2·714	1·279 2·662
Cutoff	**1·476**	**1·461**	**1·447**	**1·433**	**1·419**	**1·405**

Hit Rate	0·0700	0·0720	0·0740	0·0760	0·0780	0·0800
0·455	1·363	1·348	1·334	1·319	1·306	1·292
	2·952	2·889	2·829	2·772	2·718	2·666
0·460	1·375	1·361	1·346	1·332	1·318	1·305
	2·956	2·893	2·833	2·776	2·722	2·670
0·465	1·388	1·373	1·359	1·345	1·331	1·317
	2·960	2·896	2·836	2·779	2·725	2·673
0·470	1·401	1·386	1·371	1·357	1·343	1·330
	2·963	2·899	2·839	2·782	2·728	2·676
0·475	1·413	1·398	1·384	1·370	1·356	1·342
	2·965	2·902	2·842	2·784	2·730	2·678
0·480	1·426	1·411	1·396	1·382	1·369	1·355
	2·968	2·904	2·844	2·786	2·732	2·680
0·485	1·438	1·423	1·409	1·395	1·381	1·367
	2·969	2·906	2·845	2·788	2·733	2·682
0·490	1·451	1·436	1·422	1·407	1·394	1·380
	2·970	2·907	2·846	2·789	2·735	2·683
0·495	1·463	1·449	1·434	1·420	1·406	1·393
	2·971	2·907	2·847	2·790	2·735	2·683
0·500	1·476	1·461	1·447	1·433	1·419	1·405
	2·971	2·908	2·847	2·790	2·735	2·683
0·505	1·488	1·474	1·459	1·445	1·431	1·418
	2·971	2·907	2·847	2·790	2·735	2·683
0·510	1·501	1·486	1·472	1·458	1·444	1·430
	2·970	2·907	2·846	2·789	2·735	2·683
0·515	1·513	1·499	1·484	1·470	1·456	1·443
	2·969	2·906	2·845	2·788	2·733	2·682
0·520	1·526	1·511	1·497	1·483	1·469	1·455
	2·968	2·904	2·844	2·786	2·732	2·680
0·525	1·538	1·524	1·509	1·495	1·481	1·468
	2·965	2·902	2·842	2·784	2·730	2·678
0·530	1·551	1·536	1·522	1·508	1·494	1·480
	2·963	2·899	2·839	2·782	2·728	2·676
0·535	1·564	1·549	1·534	1·520	1·506	1·493
	2·960	2·896	2·836	2·779	2·725	2·673
0·540	1·576	1·561	1·547	1·533	1·519	1·506
	2·956	2·893	2·833	2·776	2·722	2·670
0·545	1·589	1·574	1·560	1·546	1·532	1·518
	2·952	2·889	2·829	2·772	2·718	2·666
0·550	1·601	1·587	1·572	1·558	1·544	1·531
	2·948	2·885	2·825	2·768	2·714	2·662
Cutoff	1·476	1·461	1·447	1·433	1·419	1·405

Hit Rate	0·0700	0·0720	0·0740	0·0760	0·0780	0·0800
0·555	1·614	1·599	1·585	1·571	1·557	1·543
	2·943	2·880	2·820	2·763	2·709	2·658
0·560	1·627	1·612	1·598	1·583	1·570	1·556
	2·938	2·875	2·815	2·758	2·704	2·653
0·565	1·639	1·625	1·610	1·596	1·582	1·569
	2·932	2·869	2·809	2·753	2·699	2·648
0·570	1·652	1·637	1·623	1·609	1·595	1·581
	2·925	2·863	2·803	2·747	2·693	2·642
0·575	1·665	1·650	1·636	1·622	1·608	1·594
	2·919	2·856	2·797	2·741	2·687	2·636
0·580	1·678	1·663	1·649	1·634	1·621	1·607
	2·911	2·849	2·790	2·734	2·680	2·629
0·585	1·690	1·676	1·661	1·647	1·633	1·620
	2·904	2·841	2·782	2·726	2·673	2·622
0·590	1·703	1·689	1·674	1·660	1·646	1·633
	2·895	2·833	2·775	2·719	2·666	2·615
0·595	1·716	1·701	1·687	1·673	1·659	1·645
	2·887	2·825	2·766	2·710	2·658	2·607
0·600	1·729	1·714	1·700	1·686	1·672	1·658
	2·877	2·816	2·757	2·702	2·649	2·599
0·605	1·742	1·727	1·713	1·699	1·685	1·671
	2·868	2·806	2·748	2·693	2·640	2·590
0·610	1·755	1·740	1·726	1·712	1·698	1·684
	2·858	2·796	2·738	2·683	2·631	2·581
0·615	1·768	1·753	1·739	1·725	1·711	1·697
	2·847	2·786	2·728	2·673	2·621	2·571
0·620	1·781	1·767	1·752	1·738	1·724	1·711
	2·836	2·775	2·718	2·663	2·611	2·561
0·625	1·794	1·780	1·765	1·751	1·737	1·724
	2·824	2·764	2·706	2·652	2·600	2·551
0·630	1·808	1·793	1·778	1·764	1·751	1·737
	2·812	2·752	2·695	2·641	2·589	2·540
0·635	1·821	1·806	1·792	1·778	1·764	1·750
	2·799	2·740	2·683	2·629	2·577	2·528
0·640	1·834	1·820	1·805	1·791	1·777	1·764
	2·786	2·727	2·670	2·616	2·565	2·516
0·645	1·848	1·833	1·818	1·804	1·791	1·777
	2·773	2·713	2·657	2·604	2·553	2·504
0·650	1·861	1·846	1·832	1·818	1·804	1·790
	2·759	2·700	2·644	2·590	2·540	2·491
Cutoff	**1·476**	**1·461**	**1·447**	**1·433**	**1·419**	**1·405**

Hit Rate	0·0700	0·0720	0·0740	0·0760	0·0780	0·0800
0·655	1·875 2·744	1·860 2·685	1·845 2·630	1·831 2·577	1·818 2·526	1·804 2·478
0·660	1·888 2·729	1·874 2·671	1·859 2·615	1·845 2·562	1·831 2·512	1·818 2·465
0·665	1·902 2·713	1·887 2·655	1·873 2·600	1·859 2·548	1·845 2·498	1·831 2·451
0·670	1·916 2·697	1·901 2·639	1·887 2·585	1·872 2·533	1·859 2·483	1·845 2·436
0·675	1·930 2·681	1·915 2·623	1·900 2·569	1·886 2·517	1·872 2·468	1·859 2·421
0·680	1·943 2·663	1·929 2·606	1·914 2·552	1·900 2·501	1·886 2·452	1·873 2·405
0·685	1·958 2·646	1·943 2·589	1·928 2·535	1·914 2·484	1·900 2·436	1·887 2·389
0·690	1·972 2·628	1·957 2·571	1·942 2·518	1·928 2·467	1·915 2·419	1·901 2·373
0·695	1·986 2·609	1·971 2·553	1·957 2·500	1·943 2·450	1·929 2·402	1·915 2·356
0·700	2·000 2·590	1·985 2·534	1·971 2·482	1·957 2·432	1·943 2·384	1·929 2·339
0·710	2·029 2·549	2·014 2·495	2·000 2·443	1·986 2·394	1·972 2·347	1·958 2·303
0·720	2·059 2·507	2·044 2·453	2·029 2·403	2·015 2·354	2·001 2·308	1·988 2·264
0·730	2·089 2·463	2·074 2·410	2·059 2·360	2·045 2·312	2·031 2·267	2·018 2·224
0·740	2·119 2·416	2·104 2·364	2·090 2·315	2·076 2·268	2·062 2·224	2·048 2·182
0·750	2·150 2·367	2·136 2·316	2·121 2·268	2·107 2·222	2·093 2·179	2·080 2·138
0·760	2·182 2·315	2·167 2·266	2·153 2·219	2·139 2·174	2·125 2·132	2·111 2·091
0·770	2·215 2·262	2·200 2·213	2·185 2·167	2·171 2·124	2·158 2·082	2·144 2·042
0·780	2·248 2·205	2·233 2·158	2·219 2·113	2·205 2·071	2·191 2·030	2·177 1·992
0·790	2·282 2·146	2·267 2·101	2·253 2·057	2·239 2·016	2·225 1·976	2·211 1·939
0·800	2·317 2·085	2·303 2·040	2·288 1·998	2·274 1·958	2·260 1·920	2·247 1·883
Cutoff	1·476	1·461	1·447	1·433	1·419	1·405

Hit Rate	0·0700	0·0720	0·0740	0·0760	0·0780	0·0800
0·810	2·354	2·339	2·325	2·310	2·297	2·283
	2·021	1·978	1·937	1·898	1·861	1·825
0·820	2·391	2·376	2·362	2·348	2·334	2·320
	1·954	1·912	1·873	1·835	1·799	1·765
0·830	2·430	2·415	2·401	2·387	2·373	2·359
	1·885	1·844	1·806	1·770	1·735	1·702
0·840	2·470	2·456	2·441	2·427	2·413	2·400
	1·812	1·773	1·737	1·702	1·668	1·637
0·850	2·512	2·497	2·483	2·469	2·455	2·442
	1·737	1·699	1·664	1·631	1·599	1·568
0·860	2·556	2·541	2·527	2·513	2·499	2·485
	1·658	1·622	1·589	1·557	1·526	1·497
0·870	2·602	2·587	2·573	2·559	2·545	2·531
	1·576	1·542	1·510	1·479	1·451	1·423
0·880	2·651	2·636	2·622	2·607	2·594	2·580
	1·490	1·458	1·428	1·399	1·372	1·346
0·890	2·702	2·688	2·673	2·659	2·645	2·632
	1·400	1·370	1·342	1·315	1·289	1·265
0·900	2·757	2·743	2·728	2·714	2·700	2·687
	1·307	1·279	1·253	1·227	1·203	1·180
0·910	2·817	2·802	2·787	2·773	2·759	2·746
	1·209	1·184	1·159	1·136	1·113	1·092
0·920	2·881	2·866	2·852	2·838	2·824	2·810
	1·107	1·084	1·061	1·040	1·019	1·000
0·930	2·952	2·937	2·922	2·908	2·894	2·881
	1·000	0·979	0·958	0·939	0·921	0·903
0·940	3·031	3·016	3·001	2·987	2·973	2·960
	0·887	0·868	0·850	0·833	0·817	0·801
0·950	3·121	3·106	3·091	3·077	3·064	3·050
	0·768	0·752	0·736	0·721	0·707	0·694
0·960	3·226	3·212	3·197	3·183	3·169	3·156
	0·642	0·628	0·615	0·603	0·591	0·580
0·970	3·357	3·342	3·327	3·313	3·299	3·286
	0·507	0·496	0·486	0·476	0·467	0·458
0·980	3·530	3·515	3·500	3·486	3·472	3·459
	0·361	0·353	0·346	0·339	0·332	0·326
0·990	3·802	3·787	3·773	3·759	3·745	3·731
	0·198	0·194	0·190	0·186	0·183	0·179
0·999	4·566	4·551	4·537	4·523	4·509	4·495
	0·025	0·025	0·024	0·024	0·023	0·023
Cutoff	1·476	1·461	1·447	1·433	1·419	1·405

Hit Rate	0·0820	0·0840	0·0860	0·0880	0·0900	0·0920
0·010	−0·935	−0·948	−0·961	−0·973	−0·986	−0·998
	0·176	0·173	0·170	0·167	0·164	0·161
0·020	−0·662	−0·675	−0·688	−0·701	−0·713	−0·725
	0·320	0·314	0·308	0·303	0·298	0·293
0·030	−0·489	−0·502	−0·515	−0·528	−0·540	−0·552
	0·449	0·441	0·433	0·426	0·419	0·412
0·040	−0·359	−0·372	−0·385	−0·398	−0·410	−0·422
	0·569	0·559	0·549	0·540	0·531	0·522
0·050	−0·253	−0·266	−0·279	−0·292	−0·304	−0·316
	0·681	0·669	0·657	0·646	0·635	0·625
0·060	−0·163	−0·176	−0·189	−0·202	−0·214	−0·226
	0·786	0·772	0·759	0·746	0·734	0·722
0·070	−0·084	−0·097	−0·110	−0·123	−0·135	−0·147
	0·886	0·871	0·855	0·841	0·827	0·813
0·080	−0·013	−0·026	−0·039	−0·052	−0·064	−0·077
	0·982	0·964	0·947	0·931	0·915	0·901
0·090	0·051	0·038	0·025	0·012	0·000	−0·012
	1·072	1·053	1·034	1·017	1·000	0·984
0·100	0·110	0·097	0·084	0·072	0·059	0·047
	1·159	1·138	1·118	1·099	1·081	1·063
0·110	0·165	0·152	0·139	0·127	0·114	0·102
	1·241	1·219	1·198	1·177	1·158	1·139
0·120	0·217	0·204	0·191	0·178	0·166	0·154
	1·321	1·297	1·274	1·253	1·232	1·212
0·130	0·265	0·252	0·239	0·227	0·214	0·202
	1·397	1·372	1·348	1·325	1·303	1·282
0·140	0·311	0·298	0·285	0·273	0·260	0·248
	1·470	1·443	1·418	1·394	1·371	1·348
0·150	0·355	0·342	0 329	0·317	0·304	0·292
	1·539	1·512	1·485	1·460	1·436	1·413
0·160	0·397	0·384	0·371	0·359	0·346	0·334
	1·606	1·578	1·550	1·524	1·498	1·474
0·170	0·438	0·424	0·412	0·399	0·387	0·374
	1·671	1·641	1·612	1·585	1·558	1·533
0·180	0·476	0·463	0·450	0·438	0·425	0·413
	1·732	1·701	1·672	1·643	1·616	1·590
0·190	0·514	0·501	0·488	0·475	0·463	0·451
	1·792	1·759	1·729	1·699	1·671	1·644
0·200	0·550	0·537	0·524	0·512	0·499	0·487
	1·848	1·815	1·783	1·753	1·724	1·696
Cutoff	1·392	1·379	1·366	1·353	1·341	1·329

Hit Rate	0·0820	0·0840	0·0860	0·0880	0·0900	0·0920
0·210	0·585 1·903	0·572 1·869	0·559 1·836	0·547 1·805	0·534 1·775	0·522 1·746
0·220	0·620 1·955	0·606 1·920	0·594 1·886	0·581 1·854	0·569 1·823	0·556 1·794
0·230	0·653 2·005	0·640 1·969	0·627 1·934	0·614 1·901	0·602 1·870	0·590 1·840
0·240	0·685 2·052	0·672 2·016	0·660 1·980	0·647 1·947	0·634 1·914	0·622 1·883
0·250	0·717 2·098	0·704 2·060	0·691 2·024	0·679 1·990	0·666 1·957	0·654 1·925
0·260	0·748 2·142	0·735 2·103	0·722 2·066	0·710 2·031	0·697 1·997	0·685 1·965
0·270	0·779 2·183	0·766 2·144	0·753 2·106	0·740 2·070	0·728 2·036	0·716 2·003
0·280	0·809 2·222	0·796 2·183	0·783 2·144	0·770 2·108	0·758 2·073	0·746 2·039
0·290	0·838 2·260	0·825 2·219	0·812 2·181	0·800 2·143	0·787 2·108	0·775 2·074
0·300	0·867 2·296	0·854 2·254	0·841 2·215	0·829 2·177	0·816 2·141	0·804 2·106
0·305	0·882 2·313	0·869 2·271	0·856 2·231	0·843 2·193	0·831 2·157	0·818 2·122
0·310	0·896 2·329	0·883 2·287	0·870 2·247	0·857 2·209	0·845 2·172	0·833 2·137
0·315	0·910 2·345	0·897 2·303	0·884 2·263	0·871 2·224	0·859 2·188	0·847 2·152
0·320	0·924 2·361	0·911 2·319	0·898 2·278	0·885 2·239	0·873 2·202	0·861 2·167
0·325	0·938 2·376	0·925 2·334	0·912 2·293	0·899 2·254	0·887 2·216	0·875 2·181
0·330	0·952 2·391	0·939 2·348	0·926 2·307	0·913 2·268	0·901 2·230	0·889 2·194
0·335	0·966 2·405	0·953 2·362	0·940 2·321	0·927 2·281	0·915 2·243	0·902 2·207
0·340	0·979 2·419	0·966 2·376	0·953 2·334	0·941 2·294	0·928 2·256	0·916 2·220
0·345	0·993 2·433	0·980 2·389	0·967 2·347	0·954 2·307	0·942 2·269	0·930 2·232
0·350	1·006 2·445	0·993 2·402	0·980 2·360	0·968 2·319	0·955 2·281	0·943 2·244
Cutoff	1·392	1·379	1·366	1·353	1·341	1·329

Hit Rate	0·0820	0·0840	0·0860	0·0880	0·0900	0·0920
0·355	1·020 2·458	1·007 2·414	0·994 2·372	0·981 2·331	0·969 2·293	0·957 2·255
0·360	1·033 2·470	1·020 2·426	1·007 2·383	0·995 2·343	0·982 2·304	0·970 2·267
0·365	1·047 2·482	1·034 2·437	1·021 2·394	1·008 2·354	0·996 2·315	0·983 2·277
0·370	1·060 2·493	1·047 2·448	1·034 2·405	1·021 2·364	1·009 2·325	0·997 2·287
0·375	1·073 2·504	1·060 2·459	1·047 2·416	1·035 2·374	1·022 2·335	1·010 2·297
0·380	1·086 2·514	1·073 2·469	1·060 2·426	1·048 2·384	1·035 2·345	1·023 2·307
0·385	1·099 2·524	1·086 2·478	1·073 2·435	1·061 2·394	1·048 2·354	1·036 2·316
0·390	1·112 2·533	1·099 2·488	1·086 2·444	1·074 2·403	1·061 2·363	1·049 2·324
0·395	1·125 2·542	1·112 2·496	1·099 2·453	1·087 2·411	1·074 2·371	1·062 2·333
0·400	1·138 2·551	1·125 2·505	1·112 2·461	1·100 2·419	1·087 2·379	1·075 2·341
0·405	1·151 2·559	1·138 2·513	1·125 2·469	1·113 2·427	1·100 2·387	1·088 2·348
0·410	1·164 2·567	1·151 2·521	1·138 2·476	1·126 2·434	1·113 2·394	1·101 2·355
0·415	1·177 2·574	1·164 2·528	1·151 2·483	1·138 2·441	1·126 2·401	1·114 2·362
0·420	1·190 2·581	1·177 2·534	1·164 2·490	1·151 2·448	1·139 2·407	1·127 2·368
0·425	1·203 2·587	1·190 2·541	1·177 2·496	1·164 2·454	1·152 2·413	1·139 2·374
0·430	1·215 2·593	1·202 2·547	1·189 2·502	1·177 2·460	1·164 2·419	1·152 2·380
0·435	1·228 2·599	1·215 2·552	1·202 2·508	1·190 2·465	1·177 2·424	1·165 2·385
0·440	1·241 2·604	1·228 2·557	1·215 2·513	1·202 2·470	1·190 2·429	1·178 2·390
0·445	1·253 2·609	1·240 2·562	1·228 2·517	1·215 2·474	1·202 2·433	1·190 2·394
0·450	1·266 2·613	1·253 2·566	1·240 2·521	1·228 2·478	1·215 2·437	1·203 2·398
Cutoff	1·392	1·379	1·366	1·353	1·341	1·329

FALSE POSITIVE RATE

Hit Rate	0·0820	0·0840	0·0860	0·0880	0·0900	0·0920
0·455	1·279 2·617	1·266 2·570	1·253 2·525	1·240 2·482	1·228 2·441	1·216 2·402
0·460	1·291 2·621	1·278 2·574	1·265 2·529	1·253 2·486	1·240 2·444	1·228 2·405
0·465	1·304 2·624	1·291 2·577	1·278 2·532	1·265 2·489	1·253 2·447	1·241 2·408
0·470	1·316 2·626	1·303 2·579	1·291 2·534	1·278 2·491	1·265 2·450	1·253 2·410
0·475	1·329 2·629	1·316 2·582	1·303 2·536	1·290 2·493	1·278 2·452	1·266 2·412
0·480	1·342 2·631	1·329 2·583	1·316 2·538	1·303 2·495	1·291 2·454	1·278 2·414
0·485	1·354 2·632	1·341 2·585	1·328 2·540	1·316 2·496	1·303 2·455	1·291 2·415
0·490	1·367 2·633	1·354 2·586	1·341 2·541	1·328 2·497	1·316 2·456	1·303 2·416
0·495	1·379 2·634	1·366 2·586	1·353 2·541	1·341 2·498	1·328 2·456	1·316 2·417
0·500	1·392 2·634	1·379 2·587	1·366 2·541	1·353 2·498	1·341 2·457	1·329 2·417
0·505	1·404 2·634	1·391 2·586	1·378 2·541	1·366 2·498	1·353 2·456	1·341 2·417
0·510	1·417 2·633	1·404 2·586	1·391 2·541	1·378 2·497	1·366 2·456	1·354 2·416
0·515	1·429 2·632	1·416 2·585	1·403 2·540	1·391 2·496	1·378 2·455	1·366 2·415
0·520	1·442 2·631	1·429 2·583	1·416 2·538	1·403 2·495	1·391 2·454	1·379 2·414
0·525	1·454 2·629	1·441 2·582	1·429 2·536	1·416 2·493	1·403 2·452	1·391 2·412
0·530	1·467 2·626	1·454 2·579	1·441 2·534	1·428 2·491	1·416 2·450	1·404 2·410
0·535	1·480 2·624	1·467 2·577	1·454 2·532	1·441 2·489	1·429 2·447	1·416 2·408
0·540	1·492 2·621	1·479 2·574	1·466 2·529	1·454 2·486	1·441 2·444	1·429 2·405
0·545	1·505 2·617	1·492 2·570	1·479 2·525	1·466 2·482	1·454 2·441	1·442 2·402
0·550	1·517 2·613	1·504 2·566	1·491 2·521	1·479 2·478	1·466 2·437	1·454 2·398
Cutoff	**1·392**	**1·379**	**1·366**	**1·353**	**1·341**	**1·329**

Hit Rate	0·0820	0·0840	0·0860	0·0880	0·0900	0·0920
0·555	1·530 2·609	1·517 2·562	1·504 2·517	1·491 2·474	1·479 2·433	1·467 2·394
0·560	1·543 2·604	1·530 2·557	1·517 2·513	1·504 2·470	1·492 2·429	1·480 2·390
0·565	1·555 2·599	1·542 2·552	1·529 2·508	1·517 2·465	1·504 2·424	1·492 2·385
0·570	1·568 2·593	1·555 2·547	1·542 2·502	1·530 2·460	1·517 2·419	1·505 2·380
0·575	1·581 2·587	1·568 2·541	1·555 2·496	1·542 2·454	1·530 2·413	1·518 2·374
0·580	1·594 2·581	1·581 2·534	1·568 2·490	1·555 2·448	1·543 2·407	1·530 2·368
0·585	1·606 2·574	1·593 2·528	1·581 2·483	1·568 2·441	1·555 2·401	1·543 2·362
0·590	1·619 2·567	1·606 2·521	1·593 2·476	1·581 2·434	1·568 2·394	1·556 2·355
0·595	1·632 2·559	1·619 2·513	1·606 2·469	1·594 2·427	1·581 2·387	1·569 2·348
0·600	1·645 2·551	1·632 2·505	1·619 2·461	1·607 2·419	1·594 2·379	1·582 2·341
0·605	1·658 2·542	1·645 2·496	1·632 2·453	1·619 2·411	1·607 2·371	1·595 2·333
0·610	1·671 2·533	1·658 2·488	1·645 2·444	1·632 2·403	1·620 2·363	1·608 2·324
0·615	1·684 2·524	1·671 2·478	1·658 2·435	1·646 2·394	1·633 2·354	1·621 2·316
0·620	1·697 2·514	1·684 2·469	1·671 2·426	1·659 2·384	1·646 2·345	1·634 2·307
0·625	1·710 2·504	1·697 2·459	1·684 2·416	1·672 2·374	1·659 2·335	1·647 2·297
0·630	1·724 2·493	1·711 2·448	1·698 2·405	1·685 2·364	1·673 2·325	1·660 2·287
0·635	1·737 2·482	1·724 2·437	1·711 2·394	1·698 2·354	1·686 2·315	1·674 2·277
0·640	1·750 2·470	1·737 2·426	1·724 2·383	1·712 2·343	1·699 2·304	1·687 2·267
0·645	1·764 2·458	1·751 2·414	1·738 2·372	1·725 2·331	1·713 2·293	1·700 2·255
0·650	1·777 2·445	1·764 2·402	1·751 2·360	1·738 2·319	1·726 2·281	1·714 2·244
Cutoff	1·392	1·379	1·366	1·353	1·341	1·329

Hit Rate	0·0820	0·0840	0·0860	0·0880	0·0900	0·0920
0·655	1·791	1·778	1·765	1·752	1·740	1·727
	2·433	2·389	2·347	2·307	2·269	2·232
0·660	1·804	1·791	1·778	1·766	1·753	1·741
	2·419	2·376	2·334	2·294	2·256	2·220
0·665	1·818	1·805	1·792	1·779	1·767	1·755
	2·405	2·362	2·321	2·281	2·243	2·207
0·670	1·832	1·819	1·806	1·793	1·781	1·768
	2·391	2·348	2·307	2·268	2·230	2·194
0·675	1·846	1·832	1·820	1·807	1·795	1·782
	2·376	2·334	2·293	2·254	2·216	2·181
0·680	1·859	1·846	1·834	1·821	1·808	1·796
	2·361	2·319	2·278	2·239	2·202	2·167
0·685	1·873	1·860	1·848	1·835	1·822	1·810
	2·345	2·303	2·263	2·224	2·188	2·152
0·690	1·888	1·875	1·862	1·849	1·837	1·824
	2·329	2·287	2·247	2·209	2·172	2·137
0·695	1·902	1·889	1·876	1·863	1·851	1·839
	2·313	2·271	2·231	2·193	2·157	2·122
0·700	1·916	1·903	1·890	1·878	1·865	1·853
	2·296	2·254	2·215	2·177	2·141	2·106
0·710	1·945	1·932	1·919	1·907	1·894	1·882
	2·260	2·219	2·181	2·143	2·108	2·074
0·720	1·975	1·962	1·949	1·936	1·924	1·911
	2·222	2·183	2·144	2·108	2·073	2·039
0·730	2·005	1·991	1·979	1·966	1·954	1·941
	2·183	2·144	2·106	2·070	2·036	2·003
0·740	2·035	2·022	2·009	1·997	1·984	1·972
	2·142	2·103	2·066	2·031	1·997	1·965
0·750	2·066	2·053	2·040	2·028	2·015	2·003
	2·098	2·060	2·024	1·990	1·957	1·925
0·760	2·098	2·085	2·072	2·059	2·047	2·035
	2·052	2·016	1·980	1·947	1·914	1·883
0·770	2·131	2·118	2·105	2·092	2·080	2·067
	2·005	1·969	1·934	1·901	1·870	1·840
0·780	2·164	2·151	2·138	2·125	2·113	2·101
	1·955	1·920	1·886	1·854	1·823	1·794
0·790	2·198	2·185	2·172	2·160	2·147	2·135
	1·903	1·869	1·836	1·805	1·775	1·746
0·800	2·233	2·220	2·207	2·195	2·182	2·170
	1·848	1·815	1·783	1·753	1·724	1·696
Cutoff	**1·392**	**1·379**	**1·366**	**1·353**	**1·341**	**1·329**

Hit Rate	0·0820	0·0840	0·0860	0·0880	0·0900	0·0920
0·810	2·270 1·792	2·257 1·759	2·244 1·729	2·231 1·699	2·219 1·671	2·206 1·644
0·820	2·307 1·732	2·294 1·701	2·281 1·672	2·269 1·643	2·256 1·616	2·244 1·590
0·830	2·346 1·671	2·333 1·641	2·320 1·612	2·307 1·585	2·295 1·558	2·283 1·533
0·840	2·386 1·606	2·373 1·578	2·360 1·550	2·348 1·524	2·335 1·498	2·323 1·474
0·850	2·428 1·539	2·415 1·512	2·402 1·485	2·390 1·460	2·377 1·436	2·365 1·413
0·860	2·472 1·470	2·459 1·443	2·446 1·418	2·433 1·394	2·421 1·371	2·409 1·348
0·870	2·518 1·397	2·505 1·372	2·492 1·348	2·480 1·325	2·467 1·303	2·455 1·282
0·880	2·567 1·321	2·554 1·297	2·541 1·274	2·528 1·253	2·516 1·232	2·504 1·212
0·890	2·618 1·241	2·605 1·219	2·592 1·198	2·580 1·177	2·567 1·158	2·555 1·139
0·900	2·673 1·159	2·660 1·138	2·647 1·118	2·635 1·099	2·622 1·081	2·610 1·063
0·910	2·732 1·072	2·719 1·053	2·707 1·034	2·694 1·017	2·682 1·000	2·669 0·984
0·920	2·797 0·982	2·784 0·964	2·771 0·947	2·758 0·931	2·746 0·915	2·734 0·901
0·930	2·868 0·886	2·854 0·871	2·842 0·855	2·829 0·841	2·817 0·827	2·804 0·813
0·940	2·947 0·786	2·933 0·772	2·921 0·759	2·908 0·746	2·896 0·734	2·883 0·722
0·950	3·037 0·681	3·024 0·669	3·011 0·657	2·998 0·646	2·986 0·635	2·973 0·625
0·960	3·142 0·569	3·129 0·559	3·116 0·549	3·104 0·540	3·091 0·531	3·079 0·522
0·970	3·273 0·449	3·259 0·441	3·247 0·433	3·234 0·426	3·222 0·419	3·209 0·412
0·980	3·445 0·320	3·432 0·314	3·420 0·308	3·407 0·303	3·395 0·298	3·382 0·293
0·990	3·718 0·176	3·705 0·173	3·692 0·170	3·680 0·167	3·667 0·164	3·655 0·161
0·999	4·482 0·022	4·469 0·022	4·456 0·021	4·443 0·021	4·431 0·021	4·419 0·020
Cutoff	1·392	1·379	1·366	1·353	1·341	1·329

FALSE POSITIVE RATE

Hit Rate	0·0940	0·0960	0·0980	0·1000	0·1100	0·1200
0·010	−1·010	−1·022	−1·033	−1·045	−1·100	−1·151
	0·159	0·156	0·154	0·152	0·142	0·133
0·020	−0·737	−0·749	−0·761	−0·772	−0·827	−0·879
	0·289	0·284	0·280	0·276	0·257	0·242
0·030	−0·564	−0·576	−0·588	−0·599	−0·654	−0·706
	0·406	0·399	0·393	0·388	0·362	0·340
0·040	−0·434	−0·446	−0·458	−0·469	−0·524	−0·576
	0·514	0·506	0·498	0·491	0·458	0·431
0·050	−0·328	−0·340	−0·352	−0·363	−0·418	−0·470
	0·615	0·606	0·596	0·588	0·548	0·516
0·060	−0·238	−0·250	−0·262	−0·273	−0·328	−0·380
	0·710	0·699	0·689	0·679	0·634	0·595
0·070	−0·159	−0·171	−0·183	−0·194	−0·249	−0·301
	0·801	0·788	0·776	0·765	0·714	0·671
0·080	−0·089	−0·100	−0·112	−0·124	−0·179	−0·230
	0·886	0·873	0·860	0·847	0·791	0·743
0·090	−0·024	−0·036	−0·048	−0·059	−0·114	−0·166
	0·968	0·953	0·939	0·925	0·864	0·812
0·100	0·035	0·023	0·011	0·000	−0·055	−0·107
	1·046	1·030	1·015	1·000	0·933	0·877
0·110	0·090	0·078	0·067	0·055	0·000	−0·052
	1·121	1·104	1·087	1·071	1·000	0·940
0·120	0·142	0·130	0·118	0·107	0·052	0·000
	1·193	1·174	1·157	1·140	1·064	1·000
0·130	0·190	0·178	0·167	0·155	0·100	0·049
	1·261	1·242	1·223	1·205	1·125	1·058
0·140	0·236	0·224	0·213	0·201	0·146	0·095
	1·327	1·307	1·287	1·268	1·184	1·113
0·150	0·280	0·268	0·257	0·245	0·190	0·139
	1·390	1·369	1·348	1·329	1·240	1·166
0·160	0·322	0·310	0·299	0·287	0·232	0·181
	1·451	1·429	1·407	1·386	1·294	1·216
0·170	0·362	0·351	0·339	0·327	0·272	0·221
	1·509	1·486	1·463	1·442	1·346	1·265
0·180	0·401	0·389	0·378	0·366	0·311	0·260
	1·565	1·541	1·517	1·495	1·395	1·312
0·190	0·439	0·427	0·415	0·404	0·349	0·297
	1·618	1·593	1·569	1·546	1·443	1·357
0·200	0·475	0·463	0·451	0·440	0·385	0·333
	1·669	1·644	1·619	1·595	1·489	1·400
Cutoff	1·317	1·305	1·293	1·282	1·227	1·175

Hit Rate	0·0940	0·0960	0·0980	0·1000	0·1100	0·1200
0·210	0·510	0·498	0·487	0·475	0·420	0·369
	1·719	1·692	1·667	1·642	1·533	1·441
0·220	0·544	0·532	0·521	0·509	0·454	0·403
	1·766	1·738	1·712	1·687	1·575	1·480
0·230	0·578	0·566	0·554	0·543	0·488	0·436
	1·811	1·783	1·756	1·730	1·615	1·518
0·240	0·610	0·598	0·587	0·575	0·520	0·469
	1·854	1·825	1·798	1·771	1·653	1·554
0·250	0·642	0·630	0·619	0·607	0·552	0·500
	1·895	1·866	1·838	1·811	1·690	1·589
0·260	0·673	0·661	0·650	0·638	0·583	0·532
	1·934	1·904	1·876	1·848	1·725	1·621
0·270	0·704	0·692	0·680	0·669	0·614	0·562
	1·972	1·941	1·912	1·884	1·758	1·653
0·280	0·734	0·722	0·710	0·699	0·644	0·592
	2·007	1·976	1·947	1·918	1·790	1·683
0·290	0·763	0·751	0·740	0·728	0·673	0·622
	2·041	2·010	1·980	1·950	1·820	1·711
0·300	0·792	0·780	0·769	0·757	0·702	0·651
	2·073	2·041	2·011	1·981	1·849	1·738
0·305	0·806	0·795	0·783	0·771	0·716	0·665
	2·089	2·057	2·026	1·996	1·863	1·751
0·310	0·821	0·809	0·797	0·786	0·731	0·679
	2·104	2·071	2·040	2·010	1·876	1·764
0·315	0·835	0·823	0·811	0·800	0·745	0·693
	2·118	2·086	2·054	2·024	1·889	1·776
0·320	0·849	0·837	0·825	0·814	0·759	0·707
	2·132	2·100	2·068	2·038	1·902	1·788
0·325	0·863	0·851	0·839	0·828	0·773	0·721
	2·146	2·113	2·081	2·051	1·914	1·799
0·330	0·877	0·865	0·853	0·842	0·787	0·735
	2·159	2·126	2·094	2·064	1·926	1·810
0·335	0·890	0·879	0·867	0·855	0·800	0·749
	2·172	2·139	2·107	2·076	1·937	1·821
0·340	0·904	0·892	0·881	0·869	0·814	0·763
	2·185	2·151	2·119	2·088	1·949	1·832
0·345	0·918	0·906	0·894	0·883	0·828	0·776
	2·197	2·163	2·131	2·099	1·959	1·842
0·350	0·931	0·919	0·908	0·896	0·841	0·790
	2·209	2·175	2·142	2·111	1·970	1·852
Cutoff	1·317	1·305	1·293	1·282	1·227	1·175

Hit Rate	0·0940	0·0960	0·0980	0·1000	0·1100	0·1200
0·355	0·945 2·220	0·933 2·186	0·921 2·153	0·910 2·121	0·855 1·980	0·803 1·861
0·360	0·958 2·231	0·946 2·196	0·935 2·163	0·923 2·132	0·868 1·990	0·817 1·870
0·365	0·971 2·241	0·960 2·207	0·948 2·174	0·936 2·142	0·881 1·999	0·830 1·879
0·370	0·985 2·251	0·973 2·217	0·961 2·183	0·950 2·151	0·895 2·008	0·843 1·887
0·375	0·998 2·261	0·986 2·226	0·974 2·193	0·963 2·161	0·908 2·017	0·856 1·896
0·380	1·011 2·270	0·999 2·235	0·988 2·202	0·976 2·170	0·921 2·025	0·870 1·903
0·385	1·024 2·279	1·012 2·244	1·001 2·211	0·989 2·178	0·934 2·033	0·883 1·911
0·390	1·037 2·288	1·025 2·253	1·014 2·219	1·002 2·186	0·947 2·040	0·896 1·918
0·395	1·050 2·296	1·038 2·261	1·027 2·227	1·015 2·194	0·960 2·048	0·909 1·925
0·400	1·063 2·304	1·051 2·268	1·040 2·234	1·028 2·201	0·973 2·055	0·922 1·931
0·405	1·076 2·311	1·064 2·275	1·053 2·241	1·041 2·208	0·986 2·061	0·935 1·937
0·410	1·089 2·318	1·077 2·282	1·065 2·248	1·054 2·215	0·999 2·067	0·947 1·943
0·415	1·102 2·325	1·090 2·289	1·078 2·254	1·067 2·221	1·012 2·073	0·960 1·949
0·420	1·115 2·331	1·103 2·295	1·091 2·260	1·080 2·227	1·025 2·079	0·973 1·954
0·425	1·127 2·337	1·116 2·301	1·104 2·266	1·092 2·233	1·037 2·084	0·986 1·959
0·430	1·140 2·342	1·128 2·306	1·117 2·271	1·105 2·238	1·050 2·089	0·999 1·964
0·435	1·153 2·347	1·141 2·311	1·129 2·276	1·118 2·243	1·063 2·093	1·011 1·968
0·440	1·166 2·352	1·154 2·316	1·142 2·281	1·131 2·247	1·076 2·098	1·024 1·972
0·445	1·178 2·356	1·166 2·320	1·155 2·285	1·143 2·252	1·088 2·101	1·037 1·975
0·450	1·191 2·360	1·179 2·324	1·167 2·289	1·156 2·255	1·101 2·105	1·049 1·979
Cutoff	**1·317**	**1·305**	**1·293**	**1·282**	**1·227**	**1·175**

Hit Rate	0·0940	0·0960	0·0980	0·1000	0·1100	0·1200
0·455	1·203 2·364	1·192 2·327	1·180 2·292	1·169 2·259	1·113 2·108	1·062 1·982
0·460	1·216 2·367	1·204 2·330	1·193 2·295	1·181 2·262	1·126 2·111	1·075 1·984
0·465	1·229 2·370	1·217 2·333	1·205 2·298	1·194 2·264	1·139 2·113	1·087 1·987
0·470	1·241 2·372	1·229 2·336	1·218 2·301	1·206 2·267	1·151 2·116	1·100 1·989
0·475	1·254 2·374	1·242 2·338	1·230 2·303	1·219 2·269	1·164 2·117	1·112 1·990
0·480	1·266 2·376	1·255 2·339	1·243 2·304	1·231 2·270	1·176 2·119	1·125 1·992
0·485	1·279 2·377	1·267 2·341	1·255 2·305	1·244 2·272	1·189 2·120	1·137 1·993
0·490	1·291 2·378	1·280 2·341	1·268 2·306	1·256 2·272	1·201 2·121	1·150 1·994
0·495	1·304 2·379	1·292 2·342	1·280 2·307	1·269 2·273	1·214 2·121	1·162 1·994
0·500	1·317 2·379	1·305 2·342	1·293 2·307	1·282 2·273	1·227 2·122	1·175 1·994
0·505	1·329 2·379	1·317 2·342	1·306 2·307	1·294 2·273	1·239 2·121	1·188 1·994
0·510	1·342 2·378	1·330 2·341	1·318 2·306	1·307 2·272	1·252 2·121	1·200 1·994
0·515	1·354 2·377	1·342 2·341	1·331 2·305	1·319 2·272	1·264 2·120	1·213 1·993
0·520	1·367 2·376	1·355 2·339	1·343 2·304	1·332 2·270	1·277 2·119	1·225 1·992
0·525	1·379 2·374	1·367 2·338	1·356 2·303	1·344 2·269	1·289 2·117	1·238 1·990
0·530	1·392 2·372	1·380 2·336	1·368 2·301	1·357 2·267	1·302 2·116	1·250 1·989
0·535	1·404 2·370	1·393 2·333	1·381 2·298	1·369 2·264	1·314 2·113	1·263 1·987
0·540	1·417 2·367	1·405 2·330	1·393 2·295	1·382 2·262	1·327 2·111	1·275 1·984
0·545	1·430 2·364	1·418 2·327	1·406 2·292	1·395 2·259	1·340 2·108	1·288 1·982
0·550	1·442 2·360	1·430 2·324	1·419 2·289	1·407 2·255	1·352 2·105	1·301 1·979
Cutoff	1·317	1·305	1·293	1·282	1·227	1·175

Hit Rate	0·0940	0·0960	0·0980	0·1000	0·1100	0·1200
0·555	1·455	1·443	1·431	1·420	1·365	1·313
	2·356	2·320	2·285	2·252	2·101	1·975.
0·560	1·467	1·456	1·444	1·433	1·377	1·326
	2·352	2·316	2·281	2·247	2·098	1·972
0·565	1·480	1·468	1·457	1·445	1·390	1·339
	2·347	2·311	2·276	2·243	2·093	1·968
0·570	1·493	1·481	1·469	1·458	1·403	1·351
	2·342	2·306	2·271	2·238	2·089	1·964
0·575	1·506	1·494	1·482	1·471	1·416	1·364
	2·337	2·301	2·266	2·233	2·084	1·959
0·580	1·518	1·507	1·495	1·483	1·428	1·377
	2·331	2·295	2·260	2·227	2·079	1·954
0·585	1·531	1·519	1·508	1·496	1·441	1·390
	2·325	2·289	2·254	2·221	2·073	1·949
0·590	1·544	1·532	1·521	1·509	1·454	1·403
	2·318	2·282	2·248	2·215	2·067	1·943
0·595	1·557	1·545	1·533	1·522	1·467	1·415
	2·311	2·275	2·241	2·208	2·061	1·937
0·600	1·570	1·558	1·546	1·535	1·480	1·428
	2·304	2·268	2·234	2·201	2·055	1·931
0·605	1·583	1·571	1·559	1·548	1·493	1·441
	2·296	2·261	2·227	2·194	2·048	1·925
0·610	1·596	1·584	1·572	1·561	1·506	1·454
	2·288	2·253	2·219	2·186	2·040	1·918
0·615	1·609	1·597	1·585	1·574	1·519	1·467
	2·279	2·244	2·211	2·178	2·033	1·911
0·620	1·622	1·610	1·599	1·587	1·532	1·480
	2·270	2·235	2·202	2·170	2·025	1·903
0·625	1·635	1·623	1·612	1·600	1·545	1·494
	2·261	2·226	2·193	2·161	2·017	1·896
0·630	1·648	1·637	1·625	1·613	1·558	1·507
	2·251	2·217	2·183	2·151	2·008	1·887
0·635	1·662	1·650	1·638	1·627	1·572	1·520
	2·241	2·207	2·174	2·142	1·999	1·879
0·640	1·675	1·663	1·651	1·640	1·585	1·533
	2·231	2·196	2·163	2·132	1·990	1·870
0·645	1·688	1·677	1·665	1·653	1·598	1·547
	2·220	2·186	2·153	2·121	1·980	1·861
0·650	1·702	1·690	1·678	1·667	1·612	1·560
	2·209	2·175	2·142	2·111	1·970	1·852
Cutoff	1·317	1·305	1·293	1·282	1·227	1·175

Hit Rate	0·0940	0·0960	0·0980	0·1000	0·1100	0·1200
0·655	1·715	1·704	1·692	1·680	1·625	1·574
	2·197	2·163	2·131	2·099	1·959	1·842
0·660	1·729	1·717	1·705	1·694	1·639	1·587
	2·185	2·151	2·119	2·088	1·949	1·832
0·665	1·743	1·731	1·719	1·708	1·653	1·601
	2·172	2·139	2·107	2·076	1·937	1·821
0·670	1·756	1·745	1·733	1·721	1·666	1·615
	2·159	2·126	2·094	2·064	1·926	1·810
0·675	1·770	1·758	1·747	1·735	1·680	1·629
	2·146	2·113	2·081	2·051	1·914	1·799
0·680	1·784	1·772	1·761	1·749	1·694	1·643
	2·132	2·100	2·068	2·038	1·902	1·788
0·685	1·798	1·786	1·775	1·763	1·708	1·657
	2·118	2·086	2·054	2·024	1·889	1·776
0·690	1·812	1·801	1·789	1·777	1·722	1·671
	2·104	2·071	2·040	2·010	1·876	1·764
0·695	1·827	1·815	1·803	1·792	1·737	1·685
	2·089	2·057	2·026	1·996	1·863	1·751
0·700	1·841	1·829	1·817	1·806	1·751	1·699
	2·073	2·041	2·011	1·981	1·849	1·738
0·710	1·870	1·858	1·846	1·835	1·780	1·728
	2·041	2·010	1·980	1·950	1·820	1·711
0·720	1·899	1·888	1·876	1·864	1·809	1·758
	2·007	1·976	1·947	1·918	1·790	1·683
0·730	1·929	1·917	1·906	1·894	1·839	1·788
	1·972	1·941	1·912	1·884	1·758	1·653
0·740	1·960	1·948	1·936	1·925	1·870	1·818
	1·934	1·904	1·876	1·848	1·725	1·621
0·750	1·991	1·979	1·968	1·956	1·901	1·849
	1·895	1·866	1·838	1·811	1·690	1·589
0·760	2·023	2·011	1·999	1·988	1·933	1·881
	1·854	1·825	1·798	1·771	1·653	1·554
0·770	2·055	2·044	2·032	2·020	1·965	1·914
	1·811	1·783	1·756	1·730	1·615	1·518
0·780	2·089	2·077	2·065	2·054	1·999	1·947
	1·766	1·738	1·712	1·687	1·575	1·480
0·790	2·123	2·111	2·099	2·088	2·033	1·981
	1·719	1·692	1·667	1·642	1·533	1·441
0·800	2·158	2·146	2·135	2·123	2·068	2·017
	1·669	1·644	1·619	1·595	1·489	1·400
Cutoff	1·317	1·305	1·293	1·282	1·227	1·175

Hit Rate	0·0940	0·0960	0·0980	0·1000	0·1100	0·1200
0·810	2·194	2·183	2·171	2·159	2·104	2·053
	1·618	1·593	1·569	1·546	1·443	1·357
0·820	2·232	2·220	2·208	2·197	2·142	2·090
	1·565	1·541	1·517	1·495	1·395	1·312
0·830	2·271	2·259	2·247	2·236	2·181	2·129
	1·509	1·486	1·463	1·442	1·346	1·265
0·840	2·311	2·299	2·287	2·276	2·221	2·169
	1·451	1·429	1·407	1·386	1·294	1·216
0·850	2·353	2·341	2·329	2·318	2·263	2·211
	1·390	1·369	1·348	1·329	1·240	1·166
0·860	2·397	2·385	2·373	2·362	2·307	2·255
	1·327	1·307	1·287	1·268	1·184	1·113
0·870	2·443	2·431	2·419	2·408	2·353	2·301
	1·261	1·242	1·223	1·205	1·125	1·058
0·880	2·492	2·480	2·468	2·457	2·402	2·350
	1·193	1·174	1·157	1·140	1·064	1·000
0·890	2·543	2·531	2·520	2·508	2·453	2·402
	1·121	1·104	1·087	1·071	1·000	0·940
0·900	2·598	2·586	2·575	2·563	2·508	2·457
	1·046	1·030	1·015	1·000	0·933	0·877
0·910	2·657	2·645	2·634	2·622	2·567	2·516
	0·968	0·953	0·939	0·925	0·864	0·812
0·920	2·722	2·710	2·698	2·687	2·632	2·580
	0·886	0·873	0·860	0·847	0·791	0·743
0·930	2·792	2·780	2·769	2·757	2·702	2·651
	0·801	0·788	0·776	0·765	0·714	0·671
0·940	2·871	2·859	2·848	2·836	2·781	2·730
	0·710	0·699	0·689	0·679	0·634	0·595
0·950	2·961	2·950	2·938	2·926	2·871	2·820
	0·615	0·606	0·596	0·588	0·548	0·516
0·960	3·067	3·055	3·044	3·032	2·977	2·926
	0·514	0·506	0·498	0·491	0·458	0·431
0·970	3·197	3·185	3·174	3·162	3·107	3·056
	0·406	0·399	0·393	0·388	0·362	0·340
0·980	3·370	3·358	3·347	3·335	3·280	3·229
	0·289	0·284	0·280	0·276	0·257	0·242
0·990	3·643	3·631	3·619	3·608	3·553	3·501
	0·159	0·156	0·154	0·152	0·142	0·133
0·999	4·407	4·395	4·383	4·372	4·317	4·265
	0·020	0·020	0·019	0·019	0·018	0·017
Cutoff	1·317	1·305	1·293	1·282	1·227	1·175

Hit Rate	0·1300	0·1400	0·1500	0·1600	0·1700	0·1800
0·010	-1·200	-1·246	-1·290	-1·332	-1·372	-1·411
	0·126	0·120	0·114	0·110	0·105	0·102
0·020	-0·927	-0·973	-1·017	-1·059	-1·100	-1·138
	0·229	0·218	0·208	0·199	0·191	0·185
0·030	-0·754	-0·800	-0·844	-0·886	-0·927	-0·965
	0·322	0·306	0·292	0·280	0·269	0·259
0·040	-0·624	-0·670	-0·714	-0·756	-0·797	-0·835
	0·407	0·387	0·370	0·354	0·341	0·328
0·050	-0·518	-0·565	-0·608	-0·650	-0·691	-0·729
	0·488	0·463	0·442	0·424	0·408	0·393
0·060	-0·428	-0·474	-0·518	-0·560	-0·601	-0·639
	0·563	0·535	0·511	0·490	0·471	0·454
0·070	-0·349	-0·395	-0·439	-0·481	-0·522	-0·560
	0·635	0·603	0·576	0·552	0·531	0·512
0·080	-0·279	-0·325	-0·369	-0·411	-0·451	-0·490
	0·703	0·668	0·638	0·611	0·587	0·567
0·090	-0·214	-0·260	-0·304	-0·346	-0·387	-0·425
	0·768	0·730	0·696	0·667	0·642	0·619
0·100	-0·155	-0·201	-0·245	-0·287	-0·327	-0·366
	0·830	0·788	0·753	0·721	0·694	0·669
0·110	-0·100	-0·146	-0·190	-0·232	-0·272	-0·311
	0·889	0·845	0·806	0·773	0·743	0·717
0·120	-0·049	-0·095	-0·139	-0·181	-0·221	-0·260
	0·946	0·899	0·858	0·822	0·791	0·762
0·130	0·000	-0·046	-0·090	-0·132	-0·172	-0·211
	1·000	0·950	0·907	0·869	0·836	0·806
0·140	0·046	0·000	-0·044	-0·086	-0·126	-0·165
	1·052	1·000	0·955	0·915	0·880	0·848
0·150	0·090	0·044	0·000	-0·042	-0·082	-0·121
	1·102	1·048	1·000	0·958	0·921	0·889
0·160	0·132	0·086	0·042	0·000	-0·040	-0·079
	1·150	1·093	1·044	1·000	0·962	0·927
0·170	0·172	0·126	0·082	0·040	0·000	-0·039
	1·196	1·137	1·085	1·040	1·000	0·964
0·180	0·211	0·165	0·121	0·079	0·039	0·000
	1·240	1·179	1·125	1·078	1·037	1·000
0·190	0·248	0·202	0·159	0·117	0·076	0·037
	1·283	1·219	1·164	1·115	1·072	1·034
0·200	0·285	0·239	0·195	0·153	0·113	0·074
	1·323	1·258	1·201	1·151	1·106	1·067
Cutoff	1·126	1·080	1·036	0·994	0·954	0·915

FALSE POSITIVE RATE

Hit Rate	0·1300	0·1400	0·1500	0·1600	0·1700	0·1800
0·210	0·320 1·362	0·274 1·295	0·230 1·236	0·188 1·184	0·148 1·139	0·109 1·098
0·220	0·354 1·400	0·308 1·330	0·264 1·270	0·222 1·217	0·182 1·170	0·143 1·128
0·230	0·388 1·435	0·341 1·364	0·298 1·302	0·256 1·248	0·215 1·200	0·177 1·157
0·240	0·420 1·470	0·374 1·397	0·330 1·333	0·288 1·278	0·248 1·228	0·209 1·185
0·250	0·452 1·502	0·406 1·428	0·362 1·363	0·320 1·306	0·280 1·256	0·241 1·211
0·260	0·483 1·533	0·437 1·457	0·393 1·391	0·351 1·333	0·311 1·282	0·272 1·236
0·270	0·514 1·563	0·468 1·486	0·424 1·418	0·382 1·359	0·341 1·307	0·303 1·260
0·280	0·544 1·591	0·497 1·512	0·454 1·444	0·412 1·384	0·371 1·330	0·333 1·283
0·290	0·573 1·618	0·527 1·538	0·483 1·468	0·441 1·407	0·401 1·353	0·362 1·305
0·300	0·602 1·644	0·556 1·562	0·512 1·491	0·470 1·429	0·430 1·374	0·391 1·325
0·305	0·616 1·656	0·570 1·574	0·526 1·502	0·484 1·440	0·444 1·384	0·405 1·335
0·310	0·631 1·668	0·584 1·585	0·541 1·513	0·499 1·450	0·458 1·394	0·420 1·344
0·315	0·645 1·679	0·599 1·596	0·555 1·524	0·513 1·460	0·472 1·404	0·434 1·354
0·320	0·659 1·690	0·613 1·607	0·569 1·534	0·527 1·470	0·486 1·413	0·448 1·363
0·325	0·673 1·701	0·627 1·617	0·583 1·544	0·541 1·479	0·500 1·422	0·462 1·372
0·330	0·686 1·712	0·640 1·627	0·597 1·553	0·555 1·488	0·514 1·431	0·475 1·380
0·335	0·700 1·722	0·654 1·637	0·610 1·563	0·568 1·497	0·528 1·440	0·489 1·388
0·340	0·714 1·732	0·668 1·646	0·624 1·572	0·582 1·506	0·542 1·448	0·503 1·396
0·345	0·728 1·742	0·681 1·655	0·638 1·580	0·596 1·514	0·555 1·456	0·517 1·404
0·350	0·741 1·751	0·695 1·664	0·651 1·589	0·609 1·522	0·569 1·464	0·530 1·412
Cutoff	1·126	1·080	1·036	0·994	0·954	0·915

Hit Rate	0·1300	0·1400	0·1500	0·1600	0·1700	0·1800
0·355	0·755	0·708	0·665	0·623	0·582	0·544
	1·760	1·673	1·597	1·530	1·471	1·419
0·360	0·768	0·722	0·678	0·636	0·596	0·557
	1·769	1·681	1·605	1·538	1·478	1·426
0·365	0·781	0·735	0·691	0·649	0·609	0·570
	1·777	1·689	1·612	1·545	1·485	1·432
0·370	0·795	0·748	0·705	0·663	0·622	0·584
	1·785	1·696	1·619	1·552	1·492	1·439
0·375	0·808	0·762	0·718	0·676	0·636	0·597
	1·793	1·704	1·626	1·558	1·498	1·445
0·380	0·821	0·775	0·731	0·689	0·649	0·610
	1·800	1·711	1·633	1·565	1·505	1·451
0·385	0·834	0·788	0·744	0·702	0·662	0·623
	1·807	1·717	1·639	1·571	1·511	1·457
0·390	0·847	0·801	0·757	0·715	0·675	0·636
	1·814	1·724	1·646	1·577	1·516	1·462
0·395	0·860	0·814	0·770	0·728	0·688	0·649
	1·820	1·730	1·651	1·583	1·522	1·467
0·400	0·873	0·827	0·783	0·741	0·701	0·662
	1·826	1·736	1·657	1·588	1·527	1·472
0·405	0·886	0·840	0·796	0·754	0·714	0·675
	1·832	1·741	1·662	1·593	1·532	1·477
0·410	0·899	0·853	0·809	0·767	0·727	0·688
	1·838	1·747	1·667	1·598	1·536	1·482
0·415	0·912	0·866	0·822	0·780	0·739	0·701
	1·843	1·752	1·672	1·602	1·541	1·486
0·420	0·924	0·878	0·835	0·793	0·752	0·713
	1·848	1·756	1·677	1·607	1·545	1·490
0·425	0·937	0·891	0·847	0·805	0·765	0·726
	1·852	1·761	1·681	1·611	1·549	1·493
0·430	0·950	0·904	0·860	0·818	0·778	0·739
	1·857	1·765	1·685	1·614	1·552	1·497
0·435	0·963	0·917	0·873	0·831	0·791	0·752
	1·861	1·769	1·688	1·618	1·556	1·500
0·440	0·975	0·929	0·885	0·843	0·803	0·764
	1·864	1·772	1·692	1·621	1·559	1·503
0·445	0·988	0·942	0·898	0·856	0·816	0·777
	1·868	1·775	1·695	1·624	1·562	1·506
0·450	1·001	0·955	0·911	0·869	0·829	0·790
	1·871	1·778	1·698	1·627	1·564	1·508
Cutoff	1·126	1·080	1·036	0·994	0·954	0·915

FALSE POSITIVE RATE

Hit Rate	0·1300	0·1400	0·1500	0·1600	0·1700	0·1800
0·455	1·013 1·874	0·967 1·781	0·923 1·700	0·881 1·629	0·841 1·566	0·802 1·511
0·460	1·026 1·876	0·980 1·783	0·936 1·702	0·894 1·631	0·854 1·569	0·815 1·513
0·465	1·039 1·879	0·992 1·785	0·949 1·704	0·907 1·633	0·866 1·570	0·828 1·515
0·470	1·051 1·881	1·005 1·787	0·961 1·706	0·919 1·635	0·879 1·572	0·840 1·516
0·475	1·064 1·882	1·018 1·789	0·974 1·708	0·932 1·636	0·891 1·573	0·853 1·517
0·480	1·076 1·883	1·030 1·790	0·986 1·709	0·944 1·638	0·904 1·575	0·865 1·518
0·485	1·089 1·885	1·043 1·791	0·999 1·710	0·957 1·638	0·917 1·575	0·878 1·519
0·490	1·101 1·885	1·055 1·792	1·011 1·710	0·969 1·639	0·929 1·576	0·890 1·520
0·495	1·114 1·886	1·068 1·792	1·024 1·711	0·982 1·640	0·942 1·576	0·903 1·520
0·500	1·126 1·886	1·080 1·792	1·036 1·711	0·994 1·640	0·954 1·577	0·915 1·520
0·505	1·139 1·886	1·093 1·792	1·049 1·711	1·007 1·640	0·967 1·576	0·928 1·520
0·510	1·151 1·885	1·105 1·792	1·062 1·710	1·020 1·639	0·979 1·576	0·940 1·520
0·515	1·164 1·885	1·118 1·791	1·074 1·710	1·032 1·638	0·992 1·575	0·953 1·519
0·520	1·177 1·883	1·130 1·790	1·087 1·709	1·045 1·638	1·004 1·575	0·966 1·518
0·525	1·189 1·882	1·143 1·789	1·099 1·708	1·057 1·636	1·017 1·573	0·978 1·517
0·530	1·202 1·881	1·156 1·787	1·112 1·706	1·070 1·635	1·029 1·572	0·991 1·516
0·535	1·214 1·879	1·168 1·785	1·124 1·704	1·082 1·633	1·042 1·570	1·003 1·515
0·540	1·227 1·876	1·181 1·783	1·137 1·702	1·095 1·631	1·055 1·569	1·016 1·513
0·545	1·239 1·874	1·193 1·781	1·149 1·700	1·107 1·629	1·067 1·566	1·028 1·511
0·550	1·252 1·871	1·206 1·778	1·162 1·698	1·120 1·627	1·080 1·564	1·041 1·508
Cutoff	**1·126**	**1·080**	**1·036**	**0·994**	**0·954**	**0·915**

Hit Rate	0·1300	0·1400	0·1500	0·1600	0·1700	0·1800
0·555	1·265	1·219	1·175	1·133	1·092	1·054
	1·868	1·775	1·695	1·624	1·562	1·506
0·560	1·277	1·231	1·187	1·145	1·105	1·066
	1·864	1·772	1·692	1·621	1·559	1·503
0·565	1·290	1·244	1·200	1·158	1·118	1·079
	1·861	1·769	1·688	1·618	1·556	1·500
0·570	1·303	1·257	1·213	1·171	1·131	1·092
	1·857	1·765	1·685	1·614	1·552	1·497
0·575	1·316	1·269	1·226	1·184	1·143	1·104
	1·852	1·761	1·681	1·611	1·549	1·493
0·580	1·328	1·282	1·238	1·196	1·156	1·117
	1·848	1·756	1·677	1·607	1·545	1·490
0·585	1·341	1·295	1·251	1·209	1·169	1·130
	1·843	1·752	1·672	1·602	1·541	1·486
0·590	1·354	1·308	1·264	1·222	1·182	1·143
	1·838	1·747	1·667	1·598	1·536	1·482
0·595	1·367	1·321	1·277	1·235	1·195	1·156
	1·832	1·741	1·662	1·593	1·532	1·477
0·600	1·380	1·334	1·290	1·248	1·208	1·169
	1·826	1·736	1·657	1·588	1·527	1·472
0·605	1·393	1·347	1·303	1·261	1·220	1·182
	1·820	1·730	1·651	1·583	1·522	1·467
0·610	1·406	1·360	1·316	1·274	1·233	1·195
	1·814	1·724	1·646	1·577	1·516	1·462
0·615	1·419	1·373	1·329	1·287	1·247	1·208
	1·807	1·717	1·639	1·571	1·511	1·457
0·620	1·432	1·386	1·342	1·300	1·260	1·221
	1·800	1·711	1·633	1·565	1·505	1·451
0·625	1·445	1·399	1·355	1·313	1·273	1·234
	1·793	1·704	1·626	1·558	1·498	1·445
0·630	1·458	1·412	1·368	1·326	1·286	1·247
	1·785	1·696	1·619	1·552	1·492	1·439
0·635	1·472	1·425	1·382	1·340	1·299	1·260
	1·777	1·689	1·612	1·545	1·485	1·432
0·640	1·485	1·439	1·395	1·353	1·313	1·274
	1·769	1·681	1·605	1·538	1·478	1·426
0·645	1·498	1·452	1·408	1·366	1·326	1·287
	1·760	1·673	1·597	1·530	1·471	1·419
0·650	1·512	1·466	1·422	1·380	1·339	1·301
	1·751	1·664	1·589	1·522	1·464	1·412
Cutoff	**1·126**	**1·080**	**1·036**	**0·994**	**0·954**	**0·915**

Hit Rate	0·1300	0·1400	0·1500	0·1600	0·1700	0·1800
0·655	1·525	1·479	1·435	1·393	1·353	1·314
	1·742	1·655	1·580	1·514	1·456	1·404
0·660	1·539	1·493	1·449	1·407	1·367	1·328
	1·732	1·646	1·572	1·506	1·448	1·396
0·665	1·553	1·506	1·463	1·421	1·380	1·342
	1·722	1·637	1·563	1·497	1·440	1·388
0·670	1·566	1·520	1·476	1·434	1·394	1·355
	1·712	1·627	1·553	1·488	1·431	1·380
0·675	1·580	1·534	1·490	1·448	1·408	1·369
	1·701	1·617	1·544	1·479	1·422	1·372
0·680	1·594	1·548	1·504	1·462	1·422	1·383
	1·690	1·607	1·534	1·470	1·413	1·363
0·685	1·608	1·562	1·518	1·476	1·436	1·397
	1·679	1·596	1·524	1·460	1·404	1·354
0·690	1·622	1·576	1·532	1·490	1·450	1·411
	1·668	1·585	1·513	1·450	1·394	1·344
0·695	1·636	1·590	1·547	1·505	1·464	1·425
	1·656	1·574	1·502	1·440	1·384	1·335
0·700	1·651	1·605	1·561	1·519	1·479	1·440
	1·644	1·562	1·491	1·429	1·374	1·325
0·710	1·680	1·634	1·590	1·548	1·508	1·469
	1·618	1·538	1·468	1·407	1·353	1·305
0·720	1·709	1·663	1·619	1·577	1·537	1·498
	1·591	1·512	1·444	1·384	1·330	1·283
0·730	1·739	1·693	1·649	1·607	1·567	1·528
	1·563	1·486	1·418	1·359	1·307	1·260
0·740	1·770	1·724	1·680	1·638	1·598	1·559
	1·533	1·457	1·391	1·333	1·282	1·236
0·750	1·801	1·755	1·711	1·669	1·629	1·590
	1·502	1·428	1·363	1·306	1·256	1·211
0·760	1·833	1·787	1·743	1·701	1·660	1·622
	1·470	1·397	1·333	1·278	1·228	1·185
0·770	1·865	1·819	1·775	1·733	1·693	1·654
	1·435	1·364	1·302	1·248	1·200	1·157
0·780	1·899	1·853	1·809	1·767	1·726	1·688
	1·400	1·330	1·270	1·217	1·170	1·128
0·790	1·933	1·887	1·843	1·801	1·761	1·722
	1·362	1·295	1·236	1·184	1·139	1·098
0·800	1·968	1·922	1·878	1·836	1·796	1·757
	1·323	1·258	1·201	1·151	1·106	1·067
Cutoff	1·126	1·080	1·036	0·994	0·954	0·915

Hit Rate	0·1300	0·1400	0·1500	0·1600	0·1700	0·1800
0·810	2·004	1·958	1·914	1·872	1·832	1·793
	1·283	1·219	1·164	1·115	1·072	1·034
0·820	2·042	1·996	1·952	1·910	1·870	1·831
	1·240	1·179	1·125	1·078	1·037	1·000
0·830	2·081	2·034	1·991	1·949	1·908	1·870
	1·196	1·137	1·085	1·040	1·000	0·964
0·840	2·121	2·075	2·031	1·989	1·949	1·910
	1·150	1·093	1·044	1·000	0·962	0·927
0·850	2·163	2·117	2·073	2·031	1·991	1·952
	1·102	1·048	1·000	0·958	0·921	0·889
0·860	2·207	2·161	2·117	2·075	2·034	1·996
	1·052	1·000	0·955	0·915	0·880	0·848
0·870	2·253	2·207	2·163	2·121	2·081	2·042
	1·000	0·950	0·907	0·869	0·836	0·806
0·880	2·301	2·255	2·211	2·169	2·129	2·090
	0·946	0·899	0·858	0·822	0·791	0·762
0·890	2·353	2·307	2·263	2·221	2·181	2·142
	0·889	0·845	0·806	0·773	0·743	0·717
0·900	2·408	2·362	2·318	2·276	2·236	2·197
	0·830	0·788	0·753	0·721	0·694	0·669
0·910	2·467	2·421	2·377	2·335	2·295	2·256
	0·768	0·730	0·696	0·667	0·642	0·619
0·920	2·531	2·485	2·442	2·400	2·359	2·320
	0·703	0·668	0·638	0·611	0·587	0·567
0·930	2·602	2·556	2·512	2·470	2·430	2·391
	0·635	0·603	0·576	0·552	0·531	0·512
0·940	2·681	2·635	2·591	2·549	2·509	2·470
	0·563	0·535	0·511	0·490	0·471	0·454
0·950	2·771	2·725	2·681	2·639	2·599	2·560
	0·488	0·463	0·442	0·424	0·408	0·393
0·960	2·877	2·831	2·787	2·745	2·705	2·666
	0·407	0·387	0·370	0·354	0·341	0·328
0·970	3·007	2·961	2·917	2·875	2·835	2·796
	0·322	0·306	0·292	0·280	0·269	0·259
0·980	3·180	3·134	3·090	3·048	3·008	2·969
	0·229	0·218	0·208	0·199	0·191	0·185
0·990	3·453	3·407	3·363	3·321	3·281	3·242
	0·126	0·120	0·114	0·110	0·105	0·102
0·999	4·217	4·171	4·127	4·085	4·044	4·006
	0·016	0·015	0·014	0·014	0·013	0·013
Cutoff	1·126	1·080	1·036	0·994	0·954	0·915

FALSE POSITIVE RATE

Hit Rate	0·1900	0·2000	0·2100	0·2200	0·2300	0·2400
0·010	-1·448 0·098	-1·485 0·095	-1·520 0·092	-1·554 0·090	-1·588 0·088	-1·620 0·086
0·020	-1·176 0·178	-1·212 0·173	-1·247 0·168	-1·282 0·164	-1·315 0·159	-1·347 0·156
0·030	-1·003 0·251	-1·039 0·243	-1·074 0·236	-1·109 0·230	-1·142 0·224	-1·174 0·219
0·040	-0·873 0·318	-0·909 0·308	-0·944 0·299	-0·978 0·291	-1·012 0·284	-1·044 0·277
0·050	-0·767 0·380	-0·803 0·368	-0·838 0·358	-0·873 0·348	-0·906 0·340	-0·939 0·332
0·060	-0·677 0·439	-0·713 0·425	-0·748 0·413	-0·783 0·402	-0·816 0·392	-0·848 0·383
0·070	-0·598 0·495	-0·634 0·480	-0·669 0·466	-0·704 0·453	-0·737 0·442	-0·769 0·432
0·080	-0·527 0·548	-0·563 0·531	-0·599 0·516	-0·633 0·502	-0·666 0·490	-0·699 0·478
0·090	-0·463 0·598	-0·499 0·580	-0·534 0·563	-0·569 0·548	-0·602 0·535	-0·634 0·522
0·100	-0·404 0·647	-0·440 0·627	-0·475 0·609	-0·509 0·593	-0·543 0·578	-0·575 0·565
0·110	-0·349 0·693	-0·385 0·672	-0·420 0·652	-0·454 0·635	-0·488 0·619	-0·520 0·605
0·120	-0·297 0·737	-0·333 0·715	-0·369 0·694	-0·403 0·676	-0·436 0·659	-0·469 0·643
0·130	-0·248 0·780	-0·285 0·756	-0·320 0·734	-0·354 0·714	-0·388 0·697	-0·420 0·680
0·140	-0·202 0·820	-0·239 0·795	-0·274 0·772	-0·308 0·752	-0·341 0·733	-0·374 0·716
0·150	-0·159 0·859	-0·195 0·833	-0·230 0·809	-0·264 0·787	-0·298 0·768	-0·330 0·750
0·160	-0·117 0·897	-0·153 0·869	-0·188 0·844	-0·222 0·822	-0·256 0·801	-0·288 0·783
0·170	-0·076 0·933	-0·113 0·904	-0·148 0·878	-0·182 0·855	-0·215 0·833	-0·248 0·814
0·180	-0·037 0·967	-0·074 0·937	-0·109 0·910	-0·143 0·886	-0·177 0·864	-0·209 0·844
0·190	0·000 1·000	-0·036 0·969	-0·071 0·942	-0·106 0·916	-0·139 0·894	-0·172 0·873
0·200	0·036 1·032	0·000 1·000	-0·035 0·971	-0·069 0·946	-0·103 0·922	-0·135 0·901
Cutoff	**0·878**	**0·842**	**0·806**	**0·772**	**0·739**	**0·706**

Hit Rate	0·1900	0·2000	0·2100	0·2200	0·2300	0·2400
0·210	0·071	0·035	0·000	-0·034	-0·068	-0·100
	1·062	1·029	1·000	0·973	0·949	0·927
0·220	0·106	0·069	0·034	0·000	-0·033	-0·066
	1·091	1·058	1·027	1·000	0·975	0·952
0·230	0·139	0·103	0·068	0·033	0·000	-0·033
	1·119	1·085	1·054	1·026	1·000	0·977
0·240	0·172	0·135	0·100	0·066	0·033	0·000
	1·146	1·110	1·079	1·050	1·024	1·000
0·250	0·203	0·167	0·132	0·098	0·064	0·032
	1·171	1·135	1·103	1·073	1·047	1·022
0·260	0·235	0·198	0·163	0·129	0·096	0·063
	1·195	1·159	1·125	1·095	1·068	1·043
0·270	0·265	0·229	0·194	0·159	0·126	0·093
	1·218	1·181	1·147	1·117	1·089	1·064
0·280	0·295	0·259	0·224	0·189	0·156	0·123
	1·240	1·202	1·168	1·137	1·109	1·083
0·290	0·325	0·288	0·253	0·219	0·185	0·153
	1·261	1·223	1·188	1·156	1·127	1·101
0·300	0·353	0·317	0·282	0·248	0·214	0·182
	1·281	1·242	1·206	1·174	1·145	1·118
0·305	0·368	0·332	0·296	0·262	0·229	0·196
	1·291	1·251	1·215	1·183	1·154	1·127
0·310	0·382	0·346	0·311	0·276	0·243	0·210
	1·300	1·260	1·224	1·191	1·162	1·135
0·315	0·396	0·360	0·325	0·290	0·257	0·225
	1·309	1·269	1·233	1·200	1·170	1·143
0·320	0·410	0·374	0·339	0·304	0·271	0·239
	1·318	1·277	1·241	1·208	1·178	1·150
0·325	0·424	0 388	0·353	0·318	0·285	0·253
	1·326	1·286	1·249	1·216	1·185	1 158
0·330	0·438	0·402	0·367	0·332	0·299	0·266
	1·335	1·294	1·257	1·223	1·193	1·165
0·335	0·452	0·415	0·380	0·346	0·313	0·280
	1·343	1·301	1·264	1·230	1·200	1·172
0·340	0·465	0·429	0·394	0·360	0·326	0·294
	1·350	1·309	1·271	1·237	1·207	1·179
0·345	0·479	0·443	0·408	0·373	0·340	0·307
	1·358	1·316	1·278	1·244	1·213	1·185
0·350	0·493	0·456	0·421	0·387	0·354	0·321
	1·365	1·323	1·285	1·251	1·220	1·191
Cutoff	0·878	0·842	0·806	0·772	0·739	0·706

Hit Rate	0·1900	0·2000	0·2100	0·2200	0·2300	0·2400
0·355	0·506	0·470	0·435	0·400	0·367	0·334
	1·372	1·330	1·292	1·257	1·226	1·198
0·360	0·519	0·483	0·448	0·414	0·380	0·348
	1·379	1·336	1·298	1·264	1·232	1·203
0·365	0·533	0·496	0·461	0·427	0·394	0·361
	1·385	1·343	1·304	1·269	1·238	1·209
0·370	0·546	0·510	0·475	0·440	0·407	0·374
	1·391	1·349	1·310	1·275	1·243	1·215
0·375	0·559	0·523	0·488	0·454	0·420	0·388
	1·397	1·354	1·316	1·281	1·249	1·220
0·380	0·572	0·536	0·501	0·467	0·433	0·401
	1·403	1·360	1·321	1·286	1·254	1·225
0·385	0·586	0·549	0·514	0·480	0·446	0·414
	1·409	1·365	1·326	1·291	1·259	1·230
0·390	0·599	0·562	0·527	0·493	0·460	0·427
	1·414	1·370	1·331	1·296	1·264	1·234
0·395	0·612	0·575	0·540	0·506	0·473	0·440
	1·419	1·375	1·336	1·300	1·268	1·239
0·400	0·625	0·588	0·553	0·519	0·485	0·453
	1·424	1·380	1·341	1·305	1·272	1·243
0·405	0·637	0·601	0·566	0·532	0·498	0·466
	1·428	1·384	1·345	1·309	1·276	1·247
0·410	0·650	0·614	0·579	0·545	0·511	0·479
	1·433	1·389	1·349	1·313	1·280	1·250
0·415	0·663	0·627	0·592	0·557	0·524	0·492
	1·437	1·393	1·353	1·317	1·284	1·254
0·420	0·676	0·640	0·605	0·570	0·537	0·504
	1·440	1·396	1·356	1·320	1·287	1·257
0·425	0·689	0·653	0·617	0·583	0·550	0·517
	1·444	1·400	1·360	1·323	1·291	1·261
0·430	0·702	0·665	0·630	0·596	0·562	0·530
	1·447	1·403	1·363	1·327	1·294	1·263
0·435	0·714	0·678	0·643	0·609	0·575	0·543
	1·451	1·406	1·366	1·329	1·296	1·266
0·440	0·727	0·691	0·655	0·621	0·588	0·555
	1·453	1·409	1·369	1·332	1·299	1·269
0·445	0·740	0·703	0·668	0·634	0·601	0·568
	1·456	1·411	1·371	1·335	1·301	1·271
0·450	0·752	0·716	0·681	0·647	0·613	0·581
	1·459	1·414	1·373	1·337	1·303	1·273
Cutoff	0·878	0·842	0·806	0·772	0·739	0·706

Hit Rate	0·1900	0·2000	0·2100	0·2200	0·2300	0·2400
0·455	0·765 1·461	0·729 1·416	0·693 1·375	0·659 1·339	0·626 1·305	0·593 1·275
0·460	0·777 1·463	0·741 1·418	0·706 1·377	0·672 1·341	0·638 1·307	0·606 1·277
0·465	0·790 1·464	0·754 1·420	0·719 1·379	0·684 1·342	0·651 1·309	0·618 1·278
0·470	0·803 1·466	0·766 1·421	0·731 1·380	0·697 1·344	0·664 1·310	0·631 1·280
0·475	0·815 1·467	0·779 1·422	0·744 1·382	0·709 1·345	0·676 1·311	0·644 1·281
0·480	0·828 1·468	0·791 1·423	0·756 1·383	0·722 1·346	0·689 1·312	0·656 1·282
0·485	0·840 1·469	0·804 1·424	0·769 1·383	0·735 1·346	0·701 1·313	0·669 1·282
0·490	0·853 1·470	0·817 1·425	0·781 1·384	0·747 1·347	0·714 1·313	0·681 1·283
0·495	0·865 1·470	0·829 1·425	0·794 1·384	0·760 1·347	0·726 1·314	0·694 1·283
0·500	0·878 1·470	0·842 1·425	0·806 1·384	0·772 1·347	0·739 1·314	0·706 1·283
0·505	0·890 1·470	0·854 1·425	0·819 1·384	0·785 1·347	0·751 1·314	0·719 1·283
0·510	0·903 1·470	0·867 1·425	0·831 1·384	0·797 1·347	0·764 1·313	0·731 1·283
0·515	0·916 1·469	0·879 1·424	0·844 1·383	0·810 1·346	0·776 1·313	0·744 1·282
0·520	0·928 1·468	0·892 1·423	0·857 1·383	0·822 1·346	0·789 1·312	0·756 1·282
0·525	0·941 1·467	0·904 1·422	0·869 1·382	0·835 1·345	0·802 1·311	0·769 1·281
0·530	0·953 1·466	0·917 1·421	0·882 1·380	0·847 1·344	0·814 1·310	0·782 1·280
0·535	0·966 1·464	0·929 1·420	0·894 1·379	0·860 1·342	0·827 1·309	0·794 1·278
0·540	0·978 1·463	0·942 1·418	0·907 1·377	0·873 1·341	0·839 1·307	0·807 1·277
0·545	0·991 1·461	0·955 1·416	0·919 1·375	0·885 1·339	0·852 1·305	0·819 1·275
0·550	1·004 1·459	0·967 1·414	0·932 1·373	0·898 1·337	0·865 1·303	0·832 1·273
Cutoff	0·878	0·842	0·806	0·772	0·739	0·706

FALSE POSITIVE RATE

Hit Rate	0·1900	0·2000	0·2100	0·2200	0·2300	0·2400
0·555	1·016 1·456	0·980 1·411	0·945 1·371	0·910 1·335	0·877 1·301	0·845 1·271
0·560	1·029 1·453	0·993 1·409	0·957 1·369	0·923 1·332	0·890 1·299	0·857 1·269
0·565	1·042 1·451	1·005 1·406	0·970 1·366	0·936 1·329	0·903 1·296	0·870 1·266
0·570	1·054 1·447	1·018 1·403	0·983 1·363	0·949 1·327	0·915 1·294	0·883 1·263
0·575	1·067 1·444	1·031 1·400	0·996 1·360	0·961 1·323	0·928 1·291	0·895 1·261
0·580	1·080 1·440	1·044 1·396	1·008 1·356	0·974 1·320	0·941 1·287	0·908 1·257
0·585	1·093 1·437	1·056 1·393	1·021 1·353	0·987 1·317	0·954 1·284	0·921 1·254
0·590	1·105 1·433	1·069 1·389	1·034 1·349	1·000 1·313	0·966 1·280	0·934 1·250
0·595	1·118 1·428	1·082 1·384	1·047 1·345	1·013 1·309	0·979 1·276	0·947 1·247
0·600	1·131 1·424	1·095 1·380	1·060 1·341	1·026 1·305	0·992 1·272	0·960 1·243
0·605	1·144 1·419	1·108 1·375	1·073 1·336	1·039 1·300	1·005 1·268	0·973 1·239
0·610	1·157 1·414	1·121 1·370	1·086 1·331	1·052 1·296	1·018 1·264	0·986 1·234
0·615	1·170 1·409	1·134 1·365	1·099 1·326	1·065 1·291	1·031 1·259	0·999 1·230
0·620	1·183 1·403	1·147 1·360	1·112 1·321	1·078 1·286	1·044 1·254	1·012 1·225
0·625	1·197 1·397	1·160 1·354	1·125 1·316	1·091 1·281	1·057 1·249	1·025 1·220
0·630	1·210 1·391	1·173 1·349	1·138 1·310	1·104 1·275	1·071 1·243	1·038 1·215
0·635	1·223 1·385	1·187 1·343	1·152 1·304	1·117 1·269	1·084 1·238	1·051 1·209
0·640	1·236 1·379	1·200 1·336	1·165 1·298	1·131 1·264	1·097 1·232	1·065 1·203
0·645	1·250 1·372	1·213 1·330	1·178 1·292	1·144 1·257	1·111 1·226	1·078 1·198
0·650	1·263 1·365	1·227 1·323	1·192 1·285	1·158 1·251	1·124 1·220	1·092 1·191
Cutoff	0·878	0·842	0·806	0·772	0·739	0·706

Hit Rate	0·1900	0·2000	0·2100	0·2200	0·2300	0·2400
0·655	1·277 1·358	1·240 1·316	1·205 1·278	1·171 1·244	1·138 1·213	1·105 1·185
0·660	1·290 1·350	1·254 1·309	1·219 1·271	1·185 1·237	1·151 1·207	1·119 1·179
0·665	1·304 1·343	1·268 1·301	1·233 1·264	1·198 1·230	1·165 1·200	1·132 1·172
0·670	1·318 1·335	1·282 1·294	1·246 1·257	1·212 1·223	1·179 1·193	1·146 1·165
0·675	1·332 1·326	1·295 1·286	1·260 1·249	1·226 1·216	1·193 1·185	1·160 1·158
0·680	1·346 1·318	1·309 1·277	1·274 1·241	1·240 1·208	1·207 1·178	1·174 1·150
0·685	1·360 1·309	1·323 1·269	1·288 1·233	1·254 1·200	1·221 1·170	1·188 1·143
0·690	1·374 1·300	1·337 1·260	1·302 1·224	1·268 1·191	1·235 1·162	1·202 1·135
0·695	1·388 1·291	1·352 1·251	1·316 1·215	1·282 1·183	1·249 1·154	1·216 1·127
0·700	1·402 1·281	1·366 1·242	1·331 1·206	1·297 1·174	1·263 1·145	1·231 1·118
0·710	1·431 1·261	1·395 1·223	1·360 1·188	1·326 1·156	1·292 1·127	1·260 1·101
0·720	1·461 1·240	1·424 1·202	1·389 1·168	1·355 1·137	1·322 1·109	1·289 1·083
0·730	1·491 1·218	1·454 1·181	1·419 1·147	1·385 1·117	1·352 1·089	1·319 1·064
0·740	1·521 1·195	1·485 1·159	1·450 1·125	1·416 1·095	1·382 1·068	1·350 1·043
0·750	1·552 1·171	1·516 1·135	1·481 1·103	1·447 1·073	1·413 1·047	1·381 1·022
0·760	1·584 1·146	1·548 1·110	1·513 1·079	1·478 1·050	1·445 1·024	1·413 1·000
0·770	1·617 1·119	1·580 1·085	1·545 1·054	1·511 1·026	1·478 1·000	1·445 0·977
0·780	1·650 1·091	1·614 1·058	1·579 1·027	1·544 1·000	1·511 0·975	1·478 0·952
0·790	1·684 1·062	1·648 1·029	1·613 1·000	1·579 0·973	1·545 0·949	1·513 0·927
0·800	1·720 1·032	1·683 1·000	1·648 0·971	1·614 0·946	1·580 0·922	1·548 0·901
Cutoff	0·878	0·842	0·806	0·772	0·739	0·706

Hit Rate	0·1900	0·2000	0·2100	0·2200	0·2300	0·2400
0·810	1·756	1·720	1·684	1·650	1·617	1·584
	1·000	0·969	0·942	0·916	0·894	0·873
0·820	1·793	1·757	1·722	1·688	1·654	1·622
	0·967	0·937	0·910	0·886	0·864	0·844
0·830	1·832	1·796	1·761	1·726	1·693	1·660
	0·933	0·904	0·878	0·855	0·833	0·814
0·840	1·872	1·836	1·801	1·767	1·733	1·701
	0·897	0·869	0·844	0·822	0·801	0·783
0·850	1·914	1·878	1·843	1·809	1·775	1·743
	0·859	0·833	0·809	0·787	0·768	0·750
0·860	1·958	1·922	1·887	1·853	1·819	1·787
	0·820	0·795	0·772	0·752	0·733	0·716
0·870	2·004	1·968	1·933	1·899	1·865	1·833
	0·780	0·756	0·734	0·714	0·697	0·680
0·880	2·053	2·017	1·981	1·947	1·914	1·881
	0·737	0·715	0·694	0·676	0·659	0·643
0·890	2·104	2·068	2·033	1·999	1·965	1·933
	0·693	0·672	0·652	0·635	0·619	0·605
0·900	2·159	2·123	2·088	2·054	2·020	1·988
	0·647	0·627	0·609	0·593	0·578	0·565
0·910	2·219	2·182	2·147	2·113	2·080	2·047
	0·598	0·580	0·563	0·548	0·535	0·522
0·920	2·283	2·247	2·211	2·177	2·144	2·111
	0·548	0·531	0·516	0·502	0·490	0·478
0·930	2·354	2·317	2·282	2·248	2·215	2·182
	0·495	0·480	0·466	0·453	0·442	0·432
0·940	2·433	2·396	2·361	2·327	2·294	2·261
	0·439	0·425	0·413	0·402	0·392	0·383
0·950	2·523	2·486	2·451	2·417	2·384	2·351
	0·380	0·368	0·358	0·348	0·340	0·332
0·960	2·629	2·592	2·557	2·523	2·490	2·457
	0·318	0·308	0·299	0·291	0·284	0·277
0·970	2·759	2·722	2·687	2·653	2·620	2·587
	0·251	0·243	0·236	0·230	0·224	0·219
0·980	2·932	2·895	2·860	2·826	2·793	2·760
	0·178	0·173	0·168	0·164	0·159	0·156
0·990	3·204	3·168	3·133	3·099	3·065	3·033
	0·098	0·095	0·092	0·090	0·088	0·086
0·999	3·968	3·932	3·897	3·862	3·829	3·797
	0·012	0·012	0·012	0·011	0·011	0·011
Cutoff	0·878	0·842	0·806	0·772	0·739	0·706

Hit Rate	0·2500	0·2600	0·2700	0·2800	0·2900	0·3000
0·010	−1·652	−1·683	−1·714	−1·744	−1·773	−1·802
	0·084	0·082	0·081	0·079	0·078	0·077
0·020	−1·379	−1·410	−1·441	−1·471	−1·500	−1·529
	0·152	0·149	0·146	0·144	0·141	0·139
0·030	−1·206	−1·237	−1·268	−1·298	−1·327	−1·356
	0·214	0·210	0·206	0·202	0·199	0·196
0·040	−1·076	−1·107	−1·138	−1·168	−1·197	−1·226
	0·271	0·266	0·261	0·256	0·252	0·248
0·050	−0·970	−1·002	−1·032	−1·062	−1·091	−1·120
	0·325	0·318	0·312	0·306	0·301	0·297
0·060	−0·880	−0·911	−0·942	−0·972	−1·001	−1·030
	0·375	0·367	0·360	0·354	0·348	0·343
0·070	−0·801	−0·832	−0·863	−0·893	−0·922	−0·951
	0·423	0·414	0·406	0·399	0·392	0·386
0·080	−0·731	−0·762	−0·792	−0·822	−0·852	−0·881
	0·468	0·458	0·450	0·442	0·434	0·428
0·090	−0·666	−0·697	−0·728	−0·758	−0·787	−0·816
	0·511	0·501	0·491	0·482	0·474	0·467
0·100	−0·607	−0·638	−0·669	−0·699	−0·728	−0·757
	0·552	0·541	0·531	0·521	0·513	0·505
0·110	−0·552	−0·583	−0·614	−0·644	−0·673	−0·702
	0·592	0·580	0·569	0·559	0·549	0·541
0·120	−0·500	−0·532	−0·562	−0·592	−0·622	−0·651
	0·630	0·617	0·605	0·594	0·584	0·575
0·130	−0·452	−0·483	−0·514	−0·544	−0·573	−0·602
	0·666	0·652	0·640	0·628	0·618	0·608
0·140	−0·406	−0·437	−0·468	−0·497	−0·527	−0·556
	0·700	0·686	0·673	0·661	0·650	0·640
0·150	−0·362	0·393	−0·424	−0·454	−0·483	−0·512
	0·734	0·719	0·705	0·693	0·681	0·671
0·160	0·320	−0·351	−0·382	−0·412	−0·441	−0·470
	0·766	0·750	0·736	0·723	0·711	0·700
0·170	−0·280	−0·311	−0·341	−0·371	−0·401	−0·430
	0·796	0·780	0·765	0·752	0·739	0·728
0·180	−0·241	−0·272	−0·303	−0·333	−0·362	−0·391
	0·826	0·809	0·794	0·780	0·767	0·755
0·190	−0·203	−0·235	−0·265	−0·295	−0·325	−0·353
	0·854	0·837	0·821	0·806	0·793	0·780
0·200	−0·167	−0·198	−0·229	−0·259	−0·288	−0·317
	0·881	0·863	0·847	0·832	0·818	0·805
Cutoff	0·674	0·643	0·613	0·583	0·553	0·524

Hit Rate	0·2500	0·2600	0·2700	0·2800	0·2900	0·3000
0·210	−0·132	−0·163	−0·194	−0·224	−0·253	−0·282
	0·907	0·889	0·872	0·856	0·842	0·829
0·220	−0·098	−0·129	−0·159	−0·189	−0·219	−0·248
	0·932	0·913	0·896	0·880	0·865	0·852
0·230	−0·064	−0·096	−0·126	−0·156	−0·185	−0·214
	0·956	0·936	0·918	0·902	0·887	0·873
0·240	−0·032	−0·063	−0·093	−0·123	−0·153	−0·182
	0·978	0·958	0·940	0·924	0·908	0·894
0·250	0·000	−0·031	−0·062	−0·092	−0·121	−0·150
	1·000	0·980	0·961	0·944	0·928	0·914
0·260	0·031	0·000	−0·031	−0·061	−0·090	−0·119
	1·021	1·000	0·981	0·964	0·948	0·933
0·270	0·062	0·031	0·000	−0·030	−0·059	−0·088
	1·040	1·019	1·000	0·982	0·966	0·951
0·280	0·092	0·061	0·030	0·000	−0·029	−0·058
	1·059	1·038	1·018	1·000	0·983	0·968
0·290	0·121	0·090	0·059	0·029	0·000	−0·029
	1·077	1·055	1·035	1·017	1·000	0·985
0·300	0·150	0·119	0·088	0·058	0·029	0·000
	1·094	1·072	1·052	1·033	1·016	1·000
0·305	0·164	0·133	0·103	0·073	0·043	0·014
	1·102	1·080	1·059	1·041	1·023	1·007
0·310	0·179	0·147	0·117	0·087	0·058	0·029
	1·110	1·088	1·067	1·048	1·031	1·015
0·315	0·193	0·162	0·131	0·101	0·072	0·043
	1·118	1·095	1·074	1·055	1·038	1·022
0·320	0·207	0·176	0·145	0·115	0·086	0·057
	1·125	1·102	1·082	1·062	1·045	1·029
0·325	0·221	0·190	0·159	0·129	0·100	0·071
	1·133	1·110	1·089	1·069	1·051	1·035
0·330	0·235	0·203	0·173	0·143	0·113	0·084
	1·140	1·116	1·095	1·076	1·058	1·042
0·335	0·248	0·217	0·187	0·157	0·127	0·098
	1·146	1·123	1·102	1·082	1·064	1·048
0·340	0·262	0·231	0·200	0·170	0·141	0·112
	1·153	1·130	1·108	1·088	1·070	1·054
0·345	0·276	0·244	0·214	0·184	0·155	0·126
	1·159	1·136	1·114	1·095	1·076	1·060
0·350	0·289	0·258	0·227	0·198	0·168	0·139
	1·166	1·142	1·120	1·100	1·082	1·065
Cutoff	0·674	0·643	0·613	0·583	0·553	0·524

Hit Rate	0·2500	0·2600	0·2700	0·2800	0·2900	0·3000
0·355	0·303 1·172	0·271 1·148	0·241 1·126	0·211 1·106	0·182 1·088	0·153 1·071
0·360	0·316 1·177	0·285 1·153	0·254 1·131	0·224 1·111	0·195 1·093	0·166 1·076
0·365	0·329 1·183	0·298 1·159	0·268 1·137	0·238 1·117	0·208 1·098	0·179 1·081
0·370	0·343 1·188	0·311 1·164	0·281 1·142	0·251 1·122	0·222 1·103	0·193 1·086
0·375	0·356 1·193	0·325 1·169	0·294 1·147	0·264 1·126	0·235 1·108	0·206 1·091
0·380	0·369 1·198	0·338 1·174	0·307 1·152	0·277 1·131	0·248 1·112	0·219 1·095
0·385	0·382 1·203	0·351 1·178	0·320 1·156	0·290 1·136	0·261 1·117	0·232 1·099
0·390	0·395 1·207	0·364 1·183	0·333 1·160	0·304 1·140	0·274 1·121	0·245 1·104
0·395	0·408 1·212	0·377 1·187	0·347 1·165	0·317 1·144	0·287 1·125	0·258 1·107
0·400	0·421 1·216	0·390 1·191	0·359 1·168	0·329 1·148	0·300 1·129	0·271 1·111
0·405	0·434 1·220	0·403 1·195	0·372 1·172	0·342 1·151	0·313 1·132	0·284 1·115
0·410	0·447 1·223	0·416 1·198	0·385 1·176	0·355 1·155	0·326 1·136	0·297 1·118
0·415	0·460 1·227	0·429 1·202	0·398 1·179	0·368 1·158	0·339 1·139	0·310 1·121
0·420	0·473 1·230	0·441 1·205	0·411 1·182	0·381 1·161	0·351 1·142	0·323 1·124
0·425	0·485 1·233	0·454 1·208	0·424 1·185	0·394 1·164	0·364 1·145	0·335 1·127
0·430	0·498 1·236	0·467 1·211	0·436 1·188	0·406 1·167	0·377 1·147	0·348 1·130
0·435	0·511 1·239	0·480 1·214	0·449 1·191	0·419 1·169	0·390 1·150	0·361 1·132
0·440	0·524 1·241	0·492 1·216	0·462 1·193	0·432 1·172	0·402 1·152	0·373 1·134
0·445	0·536 1·243	0·505 1·218	0·475 1·195	0·445 1·174	0·415 1·154	0·386 1·136
0·450	0·549 1·246	0·518 1·220	0·487 1·197	0·457 1·176	0·428 1·156	0·399 1·138
Cutoff	0·674	0·643	0·613	0·583	0·553	0·524

Hit Rate	0·2500	0·2600	0·2700	0·2800	0·2900	0·3000
0·455	0·561	0·530	0·500	0·470	0·440	0·411
	1·247	1·222	1·199	1·178	1·158	1·140
0·460	0·574	0·543	0·512	0·482	0·453	0·424
	1·249	1·224	1·200	1·179	1·160	1·142
0·465	0·587	0·556	0·525	0·495	0·466	0·437
	1·251	1·225	1·202	1·181	1·161	1·143
0·470	0·599	0·568	0·538	0·508	0·478	0·449
	1·252	1·226	1·203	1·182	1·162	1·144
0·475	0·612	0·581	0·550	0·520	0·491	0·462
	1·253	1·228	1·204	1·183	1·163	1·145
0·480	0·624	0·593	0·563	0·533	0·503	0·474
	1·254	1·228	1·205	1·184	1·164	1·146
0·485	0·637	0·606	0·575	0·545	0·516	0·487
	1·255	1·229	1·206	1·184	1·165	1·147
0·490	0·649	0·618	0·588	0·558	0·528	0·499
	1·255	1·230	1·206	1·185	1·165	1·147
0·495	0·662	0·631	0·600	0·570	0·541	0·512
	1·255	1·230	1·206	1·185	1·165	1·147
0·500	0·674	0·643	0·613	0·583	0·553	0·524
	1·255	1·230	1·207	1·185	1·165	1·147
0·505	0·687	0·656	0·625	0·595	0·566	0·537
	1·255	1·230	1·206	1·185	1·165	1·147
0·510	0·700	0·668	0·638	0·608	0·578	0·549
	1·255	1·230	1·206	1·185	1·165	1·147
0·515	0·712	0·681	0·650	0·620	0·591	0·562
	1·255	1·229	1·206	1·184	1·165	1·147
0·520	0·725	0·693	0·663	0·633	0·604	0·575
	1·254	1·228	1·205	1·184	1·164	1·146
0·525	0·737	0·706	0·676	0·646	0·616	0·587
	1·253	1·228	1·204	1·183	1·163	1·145
0·530	0·750	0·719	0·688	0·658	0·629	0·600
	1·252	1·226	1·203	1·182	1·162	1·144
0·535	0·762	0·731	0·701	0·671	0·641	0·612
	1·251	1·225	1·202	1·181	1·161	1·143
0·540	0·775	0·744	0·713	0·683	0·654	0·625
	1·249	1·224	1·200	1·179	1·160	1·142
0·545	0·788	0·756	0·726	0·696	0·666	0·637
	1·247	1·222	1·199	1·178	1·158	1·140
0·550	0·800	0·769	0·738	0·709	0·679	0·650
	1·246	1·220	1·197	1·176	1·156	1·138
Cutoff	**0·674**	**0·643**	**0·613**	**0·583**	**0·553**	**0·524**

Hit Rate	0·2500	0·2600	0·2700	0·2800	0·2900	0·3000
0·555	0·813	0·782	0·751	0·721	0·692	0·663
	1·243	1·218	1·195	1·174	1·154	1·136
0·560	0·825	0·794	0·764	0·734	0·704	0·675
	1·241	1·216	1·193	1·172	1·152	1·134
0·565	0·838	0·807	0·776	0·746	0·717	0·688
	1·239	1·214	1·191	1·169	1·150	1·132
0·570	0·851	0·820	0·789	0·759	0·730	0·701
	1·236	1·211	1·188	1·167	1·147	1·130
0·575	0·864	0·832	0·802	0·772	0·743	0·714
	1·233	1·208	1·185	1·164	1·145	1·127
0·580	0·876	0·845	0·815	0·785	0·755	0·726
	1·230	1·205	1·182	1·161	1·142	1·124
0·585	0·889	0·858	0·828	0·798	0·768	0·739
	1·227	1·202	1·179	1·158	1·139	1·121
0·590	0·902	0·871	0·840	0·810	0·781	0·752
	1·223	1·198	1·176	1·155	1·136	1·118
0·595	0·915	0·884	0·853	0·823	0·794	0·765
	1·220	1·195	1·172	1·151	1·132	1·115
0·600	0·928	0·897	0·866	0·836	0·807	0·778
	1·216	1·191	1·168	1·148	1·129	1·111
0·605	0·941	0·910	0·879	0·849	0·820	0·791
	1·212	1·187	1·165	1·144	1·125	1·107
0·610	0·954	0·923	0·892	0·862	0·833	0·804
	1·207	1·183	1·160	1·140	1·121	1·104
0·615	0·967	0·936	0·905	0·875	0·846	0·817
	1·203	1·178	1·156	1·136	1·117	1·099
0·620	0·980	0·949	0·918	0·888	0·859	0·830
	1·198	1·174	1·152	1·131	1·112	1·095
0·625	0·993	0·962	0·931	0·901	0·872	0·843
	1·193	1·169	1·147	1·126	1·108	1·091
0·630	1·006	0·975	0·945	0·915	0·885	0·856
	1·188	1·164	1·142	1·122	1·103	1·086
0·635	1·020	0·988	0·958	0·928	0·899	0·870
	1·183	1·159	1·137	1·117	1·098	1·081
0·640	1·033	1·002	0·971	0·941	0·912	0·883
	1·177	1·153	1·131	1·111	1·093	1·076
0·645	1·046	1·015	0·985	0·955	0·925	0·896
	1·172	1·148	1·126	1·106	1·088	1·071
0·650	1·060	1·029	0·998	0·968	0·939	0·910
	1·166	1·142	1·120	1·100	1·082	1·065
Cutoff	0·674	0·643	0·613	0·583	0·553	0·524

FALSE POSITIVE RATE

Hit Rate	0·2500	0·2600	0·2700	0·2800	0·2900	0·3000
0·655	1·073	1·042	1·012	0·982	0·952	0·923
	1·159	1·136	1·114	1·095	1·076	1·060
0·660	1·087	1·056	1·025	0·995	0·966	0·937
	1·153	1·130	1·108	1·088	1·070	1·054
0·665	1·101	1·069	1·039	1·009	0·980	0·951
	1·146	1·123	1·102	1·082	1·064	1·048
0·670	1·114	1·083	1·053	1·023	0·993	0·964
	1·140	1·116	1·095	1·076	1·058	1·042
0·675	1·128	1·097	1·067	1·037	1·007	0·978
	1·133	1·110	1·089	1·069	1·051	1·035
0·680	1·142	1·111	1·081	1·051	1·021	0·992
	1·125	1·102	1·082	1·062	1·045	1·029
0·685	1·156	1·125	1·095	1·065	1·035	1·006
	1·118	1·095	1·074	1·055	1·038	1·022
0·690	1·170	1·139	1·109	1·079	1·049	1·020
	1·110	1·088	1·067	1·048	1·031	1·015
0·695	1·185	1·153	1·123	1·093	1·063	1·034
	1·102	1·080	1·059	1·041	1·023	1·007
0·700	1·199	1·168	1·137	1·107	1·078	1·049
	1·094	1·072	1·052	1·033	1·016	1·000
0·710	1·228	1·197	1·166	1·136	1·107	1·078
	1·077	1·055	1·035	1·017	1·000	0·985
0·720	1·257	1·226	1·196	1·166	1·136	1·107
	1·059	1·038	1·018	1·000	0·983	0·968
0·730	1·287	1·256	1·226	1·196	1·166	1·137
	1·040	1·019	1·000	0·982	0·966	0·951
0·740	1·318	1·287	1·256	1·226	1·197	1·168
	1·021	1·000	0·981	0·964	0·948	0·933
0·750	1·349	1·318	1·287	1·257	1·228	1·199
	1·000	0·980	0·961	0·944	0·928	0·914
0·760	1·381	1·350	1·319	1·289	1·260	1·231
	0·978	0·958	0·940	0·924	0·908	0·894
0·770	1·413	1·382	1·352	1·322	1·292	1·263
	0·956	0·936	0·918	0·902	0·887	0·873
0·780	1·447	1·416	1·385	1·355	1·326	1·297
	0·932	0·913	0·896	0·880	0·865	0·852
0·790	1·481	1·450	1·419	1·389	1·360	1·331
	0·907	0·889	0·872	0·856	0·842	0·829
0·800	1·516	1·485	1·454	1·424	1·395	1·366
	0·881	0·863	0·847	0·832	0·818	0·805
Cutoff	0·674	0·643	0·613	0·583	0·553	0·524

Hit Rate	0·2500	0·2600	0·2700	0·2800	0·2900	0·3000
0·810	1·552 0·854	1·521 0·837	1·491 0·821	1·461 0·806	1·431 0·793	1·402 0·780
0·820	1·590 0·826	1·559 0·809	1·528 0·794	1·498 0·780	1·469 0·767	1·440 0·755
0·830	1·629 0·796	1·598 0·780	1·567 0·765	1·537 0·752	1·508 0·739	1·479 0·728
0·840	1·669 0·766	1·638 0·750	1·607 0·736	1·577 0·723	1·548 0·711	1·519 0·700
0·850	1·711 0·734	1·680 0·719	1·649 0·705	1·619 0·693	1·590 0·681	1·561 0·671
0·860	1·755 0·700	1·724 0·686	1·693 0·673	1·663 0·661	1·634 0·650	1·605 0·640
0·870	1·801 0·666	1·770 0·652	1·739 0·640	1·709 0·628	1·680 0·618	1·651 0·608
0·880	1·849 0·630	1·818 0·617	1·788 0·605	1·758 0·594	1·728 0·584	1·699 0·575
0·890	1·901 0·592	1·870 0·580	1·839 0·569	1·809 0·559	1·780 0·549	1·751 0·541
0·900	1·956 0·552	1·925 0·541	1·894 0·531	1·864 0·521	1·835 0·513	1·806 0·505
0·910	2·015 0·511	1·984 0·501	1·954 0·491	1·924 0·482	1·894 0·474	1·865 0·467
0·920	2·080 0·468	2·048 0·458	2·018 0·450	1·988 0·442	1·958 0·434	1·929 0·428
0·930	2·150 0·423	2·119 0·414	2·089 0·406	2·059 0·399	2·029 0·392	2·000 0·386
0·940	2·229 0·375	2·198 0·367	2·168 0·360	2·138 0·354	2·108 0·348	2·079 0·343
0·950	2·319 0·325	2·288 0·318	2·258 0·312	2·228 0·306	2·198 0·301	2·169 0·297
0·960	2·425 0·271	2·394 0·266	2·363 0·261	2·334 0·256	2·304 0·252	2·275 0·248
0·970	2·555 0·214	2·524 0·210	2·494 0·206	2·464 0·202	2·434 0·199	2·405 0·196
0·980	2·728 0·152	2·697 0·149	2·667 0·146	2·637 0·144	2·607 0·141	2·578 0·139
0·990	3·001 0·084	2·970 0·082	2·939 0·081	2·909 0·079	2·880 0·078	2·851 0·077
0·999	3·765 0·011	3·734 0·010	3·703 0·010	3·673 0·010	3·644 0·010	3·615 0·010
Cutoff	0·674	0·643	0·613	0·583	0·553	0·524

FALSE POSITIVE RATE

Hit Rate	0·3500	0·4000	0·5000	0·6000	0·7000	0·8000
0·010	-1·941	-2·073	-2·326	-2·580	-2·851	-3·168
	0·072	0·069	0·067	0·069	0·077	0·095
0·020	-1·668	-1·800	-2·054	-2·307	-2·578	-2·895
	0·131	0·125	0·121	0·125	0·139	0·173
0·030	-1·495	-1·627	-1·881	-2·134	-2·405	-2·722
	0·184	0·176	0·171	0·176	0·196	0·243
0·040	-1·365	-1·497	-1·751	-2·004	-2·275	-2·592
	0·233	0·223	0·216	0·223	0·248	0·308
0·050	-1·260	-1·392	-1·645	-1·898	-2·169	-2·486
	0·278	0·267	0·259	0·267	0·297	0·368
0·060	-1·169	-1·301	-1·555	-1·808	-2·079	-2·396
	0·322	0·308	0·299	0·308	0·343	0·425
0·070	-1·090	-1·222	-1·476	-1·729	-2·000	-2·317
	0·362	0·348	0·337	0·348	0·386	0·480
0·080	-1·020	-1·152	-1·405	-1·658	-1·929	-2·247
	0·401	0·385	0·373	0·385	0·428	0·531
0·090	-0·955	-1·087	-1·341	-1·594	-1·865	-2·182
	0·438	0·420	0·407	0·420	0·467	0·580
0·100	-0·896	-1·028	-1·282	-1·535	-1·806	-2·123
	0·474	0·454	0·440	0·454	0·505	0·627
0·110	-0·841	-0·973	-1·227	-1·480	-1·751	-2·068
	0·508	0·487	0·471	0·487	0·541	0·672
0·120	-0·790	-0·922	-1·175	-1·428	-1·699	-2·017
	0·540	0·518	0·501	0·518	0·575	0·715
0·130	-0·741	-0·873	-1·126	-1·380	-1·651	-1·968
	0·571	0·548	0·530	0·548	0·608	0·756
0·140	-0·695	-0·827	-1·080	-1·334	-1·605	-1·922
	0·601	0·576	0·558	0·576	0·640	0·795
0·150	-0·651	-0·783	-1·036	-1·290	-1·561	-1·878
	0·629	0·604	0·584	0·604	0·671	0·833
0·160	-0·609	-0·741	-0·994	-1·248	-1·519	-1·836
	0·657	0·630	0·610	0·630	0·700	0·869
0·170	-0·569	-0·701	-0·954	-1·208	-1·479	-1·796
	0·683	0·655	0·634	0·655	0·728	0·904
0·180	-0·530	-0·662	-0·915	-1·169	-1·440	-1·757
	0·708	0·679	0·658	0·679	0·755	0·937
0·190	-0·493	-0·625	-0·878	-1·131	-1·402	-1·720
	0·733	0·702	0·680	0·702	0·780	0·969
0·200	-0·456	-0·588	-0·842	-1·095	-1·366	-1·683
	0·756	0·725	0·702	0·725	0·805	1·000
Cutoff	0·385	0·253	0·000	-0·253	-0·524	-0·842

Hit Rate	0·3500	0·4000	0·5000	0·6000	0·7000	0·8000
0·210	-0·421	-0·553	-0·806	-1·060	-1·331	-1·648
	0·778	0·746	0·722	0·746	0·829	1·029
0·220	-0·387	-0·519	-0·772	-1·026	-1·297	-1·614
	0·799	0·766	0·742	0·766	0·852	1·058
0·230	-0·354	-0·485	-0·739	-0·992	-1·263	-1·580
	0·820	0·786	0·761	0·786	0·873	1·085
0·240	-0·321	-0·453	-0·706	-0·960	-1·231	-1·548
	0·839	0·805	0·779	0·805	0·894	1·110
0·250	-0·289	-0·421	-0·674	-0·928	-1·199	-1·516
	0·858	0·823	0·797	0·823	0·914	1·135
0·260	-0·258	-0·390	-0·643	-0·897	-1·168	-1·485
	0·876	0·840	0·813	0·840	0·933	1·159
0·270	-0·227	-0·359	-0·613	-0·866	-1·137	-1·454
	0·893	0·856	0·829	0·856	0·951	1·181
0·280	-0·198	-0·329	-0·583	-0·836	-1·107	-1·424
	0·909	0·871	0·844	0·871	0·968	1·202
0·290	-0·168	-0·300	-0·553	-0·807	-1·078	-1·395
	0·924	0·886	0·858	0·886	0·985	1·223
0·300	-0·139	-0·271	-0·524	-0·778	-1·049	-1·366
	0·939	0·900	0·872	0·900	1·000	1·242
0·305	-0·125	-0·257	-0·510	-0·763	-1·034	-1·352
	0·946	0·907	0·878	0·907	1·007	1·251
0·310	-0·111	-0·243	-0·496	-0·749	-1·020	-1·337
	0·952	0·913	0·884	0·913	1·015	1·260
0·315	-0·096	-0·228	-0·482	-0·735	-1·006	-1·323
	0·959	0·919	0·890	0·919	1·022	1·269
0·320	-0·082	-0·214	-0·468	-0·721	-0·992	-1·309
	0·965	0·926	0·896	0·926	1·029	1·277
0·325	-0·068	-0·200	0·454	-0·707	-0·978	-1·295
	0·972	0·932	0·902	0·932	1·035	1·286
0·330	-0·055	-0·187	-0·440	-0·693	-0·964	-1·282
	0·978	0·937	0·908	0·937	1·042	1·294
0·335	-0·041	-0·173	-0·426	-0·679	-0·951	-1·268
	0·984	0·943	0·913	0·943	1·048	1·301
0·340	-0·027	-0·159	-0·412	-0·666	-0·937	-1·254
	0·989	0·948	0·918	0·948	1·054	1·309
0·345	-0·014	-0·146	-0·399	-0·652	-0·923	-1·240
	0·995	0·954	0·924	0·954	1·060	1·316
0·350	0·000	-0·132	-0·385	-0·639	-0·910	-1·227
	1·000	0·959	0·928	0·959	1·065	1·323
Cutoff	0·385	0·253	0·000	-0·253	-0·524	-0·842

FALSE POSITIVE RATE

Hit Rate	0·3500	0·4000	0·5000	0·6000	0·7000	0·8000
0·355	0·013 1·005	−0·119 0·964	−0·372 0·933	−0·625 0·964	−0·896 1·071	−1·213 1·330
0·360	0·027 1·010	−0·105 0·968	−0·358 0·938	−0·612 0·968	−0·883 1·076	−1·200 1·336
0·365	0·040 1·015	−0·092 0·973	−0·345 0·942	−0·598 0·973	−0·870 1·081	−1·187 1·343
0·370	0·053 1·019	−0·079 0·977	−0·332 0·946	−0·585 0·977	−0·856 1·086	−1·173 1·349
0·375	0·067 1·024	−0·065 0·982	−0·319 0·951	−0·572 0·982	−0·843 1·091	−1·160 1·354
0·380	0·080 1·028	−0·052 0·986	−0·305 0·954	−0·559 0·986	−0·830 1·095	−1·147 1·360
0·385	0·093 1·032	−0·039 0·989	−0·292 0·958	−0·546 0·989	−0·817 1·099	−1·134 1·365
0·390	0·106 1·036	−0·026 0·993	−0·279 0·962	−0·533 0·993	−0·804 1·104	−1·121 1·370
0·395	0·119 1·040	−0·013 0·997	−0·266 0·965	−0·520 0·997	−0·791 1·107	−1·108 1·375
0·400	0·132 1·043	0·000 1·000	−0·253 0·968	−0·507 1·000	−0·778 1·111	−1·095 1·380
0·405	0·145 1·046	0·013 1·003	−0·240 0·972	−0·494 1·003	−0·765 1·115	−1·082 1·384
0·410	0·158 1·050	0·026 1·006	−0·228 0·974	−0·481 1·006	−0·752 1·118	−1·069 1·389
0·415	0·171 1·053	0·039 1·009	−0·215 0·977	−0·468 1·009	−0·739 1·121	−1·056 1·393
0·420	0·183 1·055	0·051 1·012	−0·202 0·980	−0·455 1·012	−0·726 1·124	−1·044 1·396
0·425	0·196 1·058	0·064 1·014	−0·189 0·982	−0·442 1·014	−0·714 1·127	−1·031 1·400
0·430	0·209 1·060	0·077 1·017	−0·176 0·985	−0·430 1·017	−0·701 1·130	−1·018 1·403
0·435	0·222 1·063	0·090 1·019	−0·164 0·987	−0·417 1·019	−0·688 1·132	−1·005 1·406
0·440	0·234 1·065	0·102 1·021	−0·151 0·989	−0·404 1·021	−0·675 1·134	−0·993 1·409
0·445	0·247 1·067	0·115 1·023	−0·138 0·990	−0·392 1·023	−0·663 1·136	−0·980 1·411
0·450	0·260 1·069	0·128 1·024	−0·126 0·992	−0·379 1·024	−0·650 1·138	−0·967 1·414
Cutoff	0·385	0·253	0·000	−0·253	−0·524	−0·842

Hit Rate	0·3500	0·4000	0·5000	0·6000	0·7000	0·8000
0·455	0·272 1·070	0·140 1·026	-0·113 0·994	-0·366 1·026	-0·637 1·140	-0·955 1·416
0·460	0·285 1·072	0·153 1·027	-0·100 0·995	-0·354 1·027	-0·625 1·142	-0·942 1·418
0·465	0·297 1·073	0·166 1·029	-0·088 0·996	-0·341 1·029	-0·612 1·143	-0·929 1·420
0·470	0·310 1·074	0·178 1·030	-0·075 0·997	-0·329 1·030	-0·600 1·144	-0·917 1·421
0·475	0·323 1·075	0·191 1·031	-0·063 0·998	-0·316 1·031	-0·587 1·145	-0·904 1·422
0·480	0·335 1·076	0·203 1·031	-0·050 0·999	-0·304 1·031	-0·575 1·146	-0·892 1·423
0·485	0·348 1·076	0·216 1·032	-0·038 0·999	-0·291 1·032	-0·562 1·147	-0·879 1·424
0·490	0·360 1·077	0·228 1·032	-0·025 1·000	-0·278 1·032	-0·549 1·147	-0·867 1·425
0·495	0·373 1·077	0·241 1·033	-0·013 1·000	-0·266 1·033	-0·537 1·147	-0·854 1·425
0·500	0·385 1·077	0·253 1·033	0·000 1·000	-0·253 1·033	-0·524 1·147	-0·842 1·425
0·505	0·398 1·077	0·266 1·033	0·013 1·000	-0·241 1·033	-0·512 1·147	-0·829 1·425
0·510	0·410 1·077	0·278 1·032	0·025 1·000	-0·228 1·032	-0·499 1·147	-0·817 1·425
0·515	0·423 1·076	0·291 1·032	0·038 0·999	-0·216 1·032	-0·487 1·147	-0·804 1·424
0·520	0·435 1·076	0·304 1·031	0·050 0·999	-0·203 1·031	-0·474 1·146	-0·791 1·423
0·525	0·448 1·075	0·316 1·031	0·063 0·998	-0·191 1·031	-0·462 1·145	-0·779 1·422
0·530	0·461 1·074	0·329 1·030	0·075 0·997	-0·178 1·030	-0·449 1·144	-0·766 1·421
0·535	0·473 1·073	0·341 1·029	0·088 0·996	-0·166 1·029	-0·437 1·143	-0·754 1·420
0·540	0·486 1·072	0·354 1·027	0·100 0·995	-0·153 1·027	-0·424 1·142	-0·741 1·418
0·545	0·498 1·070	0·366 1·026	0·113 0·994	-0·140 1·026	-0·411 1·140	-0·729 1·416
0·550	0·511 1·069	0·379 1·024	0·126 0·992	-0·128 1·024	-0·399 1·138	-0·716 1·414
Cutoff	**0·385**	**0·253**	**0·000**	**-0·253**	**-0·524**	**-0·842**

FALSE POSITIVE RATE

Hit Rate	0·3500	0·4000	0·5000	0·6000	0·7000	0·8000
0·555	0·524	0·392	0·138	−0·115	−0·386	−0·703
	1·067	1·023	0·990	1·023	1·136	1·411
0·560	0·536	0·404	0·151	−0·102	−0·373	−0·691
	1·065	1·021	0·989	1·021	1·134	1·409
0·565	0·549	0·417	0·164	−0·090	−0·361	−0·678
	1·063	1·019	0·987	1·019	1·132	1·406
0·570	0·562	0·430	0·176	−0·077	−0·348	−0·665
	1·060	1·017	0·985	1·017	1·130	1·403
0·575	0·574	0·442	0·189	−0·064	−0·335	−0·653
	1·058	1·014	0·982	1·014	1·127	1·400
0·580	0·587	0·455	0·202	−0·051	−0·323	−0·640
	1·055	1·012	0·980	1·012	1·124	1·396
0·585	0·600	0·468	0·215	−0·039	−0·310	−0·627
	1·053	1·009	0·977	1·009	1·121	1·393
0·590	0·613	0·481	0·228	−0·026	−0·297	−0·614
	1·050	1·006	0·974	1·006	1·118	1·389
0·595	0·626	0·494	0·240	−0·013	−0·284	−0·601
	1·046	1·003	0·972	1·003	1·115	1·384
0·600	0·639	0·507	0·253	0·000	−0·271	−0·588
	1·043	1·000	0·968	1·000	1·111	1·380
0·605	0·652	0·520	0·266	0·013	−0·258	−0·575
	1·040	0·997	0·965	0·997	1·107	1·375
0·610	0·665	0·533	0·279	0·026	−0·245	−0·562
	1·036	0·993	0·962	0·993	1·104	1·370
0·615	0·678	0·546	0·292	0·039	−0·232	−0·549
	1·032	0·989	0·958	0·989	1·099	1·365
0·620	0·691	0·559	0·305	0·052	−0·219	−0·536
	1·028	0·986	0·954	0·986	1·095	1·360
0·625	0·704	0·572	0·319	0·065	−0·206	−0·523
	1·024	0·982	0·951	0·982	1·091	1·354
0·630	0·717	0·585	0·332	0·079	−0·193	−0·510
	1·019	0·977	0·946	0·977	1·086	1·349
0·635	0·730	0·598	0·345	0·092	−0·179	−0·496
	1·015	0·973	0·942	0·973	1·081	1·343
0·640	0·744	0·612	0·358	0·105	−0·166	−0·483
	1·010	0·968	0·938	0·968	1·076	1·336
0·645	0·757	0·625	0·372	0·119	−0·153	−0·470
	1·005	0·964	0·933	0·964	1·071	1·330
0·650	0·771	0·639	0·385	0·132	−0·139	−0·456
	1·000	0·959	0·928	0·959	1·065	1·323
Cutoff	0·385	0·253	0·000	−0·253	−0·524	−0·842

Hit Rate	0·3500	0·4000	0·5000	0·6000	0·7000	0·8000
0·655	0·784	0·652	0·399	0·146	−0·126	−0·443
	0·995	0·954	0·924	0·954	1·060	1·316
0·660	0·798	0·666	0·412	0·159	−0·112	−0·429
	0·989	0·948	0·918	0·948	1·054	1·309
0·665	0·811	0·679	0·426	0·173	−0·098	−0·415
	0·984	0·943	0·913	0·943	1·048	1·301
0·670	0·825	0·693	0·440	0·187	−0·084	−0·402
	0·978	0·937	0·908	0·937	1·042	1·294
0·675	0·839	0·707	0·454	0·200	−0·071	−0·388
	0·972	0·932	0·902	0·932	1·035	1·286
0·680	0·853	0·721	0·468	0·214	−0·057	−0·374
	0·965	0·926	0·896	0·926	1·029	1·277
0·685	0·867	0·735	0·482	0·228	−0·043	−0·360
	0·959	0·919	0·890	0·919	1·022	1·269
0·690	0·881	0·749	0·496	0·243	−0·029	−0·346
	0·952	0·913	0·884	0·913	1·015	1·260
0·695	0·895	0·763	0·510	0·257	−0·014	−0·332
	0·946	0·907	0·878	0·907	1·007	1·251
0·700	0·910	0·778	0·524	0·271	0·000	−0·317
	0·939	0·900	0·872	0·900	1·000	1·242
0·710	0·939	0·807	0·553	0·300	0·029	−0·288
	0·924	0·886	0·858	0·886	0·985	1·223
0·720	0·968	0·836	0·583	0·329	0·058	−0·259
	0·909	0·871	0·844	0·871	0·968	1·202
0·730	0·998	0·866	0·613	0·359	0·088	−0·229
	0·893	0·856	0·829	0·856	0·951	1·181
0·740	1·029	0·897	0·643	0·390	0·119	−0·198
	0·876	0·840	0·813	0·840	0·933	1·159
0·750	1·060	0·928	0·674	0·421	0·150	−0·167
	0·858	0·823	0·797	0·823	0·914	1·135
0·760	1·092	0·960	0·706	0·453	0·182	−0·135
	0·839	0·805	0·779	0·805	0·894	1·110
0·770	1·124	0·992	0·739	0·485	0·214	−0·103
	0·820	0·786	0·761	0·786	0·873	1·085
0·780	1·158	1·026	0·772	0·519	0·248	−0·069
	0·799	0·766	0·742	0·766	0·852	1·058
0·790	1·192	1·060	0·806	0·553	0·282	−0·035
	0·778	0·746	0·722	0·746	0·829	1·029
0·800	1·227	1·095	0·842	0·588	0·317	0·000
	0·756	0·725	0·702	0·725	0·805	1·000
Cutoff	0·385	0·253	0·000	−0·253	−0·524	−0·842

FALSE POSITIVE RATE

Hit Rate	0·3500	0·4000	0·5000	0·6000	0·7000	0·8000
0·810	1·263 0·733	1·131 0·702	0·878 0·680	0·625 0·702	0·353 0·780	0·036 0·969
0·820	1·301 0·708	1·169 0·679	0·915 0·658	0·662 0·679	0·391 0·755	0·074 0·937
0·830	1·339 0·683	1·208 0·655	0·954 0·634	0·701 0·655	0·430 0·728	0·113 0·904
0·840	1·380 0·657	1·248 0·630	0·994 0·610	0·741 0·630	0·470 0·700	0·153 0·869
0·850	1·422 0·629	1·290 0·604	1·036 0·584	0·783 0·604	0·512 0·671	0·195 0·833
0·860	1·466 0·601	1·334 0·576	1·080 0·558	0·827 0·576	0·556 0·640	0·239 0·795
0·870	1·512 0·571	1·380 0·548	1·126 0·530	0·873 0·548	0·602 0·608	0·285 0·756
0·880	1·560 0·540	1·428 0·518	1·175 0·501	0·922 0·518	0·651 0·575	0·333 0·715
0·890	1·612 0·508	1·480 0·487	1·227 0·471	0·973 0·487	0·702 0·541	0·385 0·672
0·900	1·667 0·474	1·535 0·454	1·282 0·440	1·028 0·454	0·757 0·505	0·440 0·627
0·910	1·726 0·438	1·594 0·420	1·341 0·407	1·087 0·420	0·816 0·467	0·499 0·580
0·920	1·790 0·401	1·658 0·385	1·405 0·373	1·152 0·385	0·881 0·428	0·563 0·531
0·930	1·861 0·362	1·729 0·348	1·476 0·337	1·222 0·348	0·951 0·386	0·634 0·480
0·940	1·940 0·322	1·808 0·308	1·555 0·299	1·301 0·308	1·030 0·343	0·713 0·425
0·950	2·030 0·278	1·898 0·267	1·645 0·259	1·392 0·267	1·120 0·297	0·803 0·368
0·960	2·136 0·233	2·004 0·223	1·751 0·216	1·497 0·223	1·226 0·248	0·909 0·308
0·970	2·266 0·184	2·134 0·176	1·881 0·171	1·627 0·176	1·356 0·196	1·039 0·243
0·980	2·439 0·131	2·307 0·125	2·054 0·121	1·800 0·125	1·529 0·139	1·212 0·173
0·990	2·712 0·072	2·580 0·069	2·326 0·067	2·073 0·069	1·802 0·077	1·485 0·095
0·999	3·476 0·009	3·344 0·009	3·090 0·008	2·837 0·009	2·566 0·010	2·249 0·012
Cutoff	0·385	0·253	0·000	-0·253	-0·524	-0·842

For EU product safety concerns, contact us at Calle de José Abascal, 56–1°, 28003 Madrid, Spain or eugpsr@cambridge.org.

www.ingramcontent.com/pod-product-compliance
Ingram Content Group UK Ltd.
Pitfield, Milton Keynes, MK11 3LW, UK
UKHW010854090126
466816UK00011B/244